Aerothermal Analysis of an Aeroengine Annular Combustor Concept with Angular Air Supply

Zur Erlangung des akademischen Grades eines

Doktors der Ingenieurwissenschaften

von der Fakultät für Maschinenbau des
Karlsruher Instituts für Technologie

genehmigte

Dissertation

von

Dipl.-Ing. Behdad Ariatabar

Tag der mündlichen Prüfung: 26.07.2019

Hauptreferent: Prof. Dr.-Ing. Hans-Jörg. Bauer

Korreferent: Prof. Dr.-Ing. Nikolaos Zarzalis

Forschungsberichte aus dem
Institut für Thermische Strömungsmaschinen

herausgegeben von:
Prof. Dr.-Ing. Hans-Jörg Bauer,
Lehrstuhl und Institut für Thermische Strömungsmaschinen
Karlsruher Institut für Technologie (KIT)
Kaiserstr. 12
D-76131 Karlsruhe

Bibliografische Information der Deutschen Nationalbibliothek

Die Deutsche Nationalbibliothek verzeichnet diese Publikation in der
Deutschen Nationalbibliografie; detaillierte bibliografische Daten sind
im Internet über http://dnb.d-nb.de abrufbar.

ISSN 1615-4983
ISBN 978-3-8325-4998-5

Logos Verlag Berlin GmbH
Comeniushof, Gubener Str. 47,
10243 Berlin
Tel.: +49 030 42 85 10 90
Fax: +49 030 42 85 10 92
INTERNET: http://www.logos-verlag.de

Vorwort des Herausgebers

Der schnelle technische Fortschritt im Turbomaschinenbau, der durch extreme technische Forderungen und starken internationalen Wettbewerb geprägt ist, verlangt einen effizienten Austausch und die Diskussion von Fachwissen und Erfahrung zwischen Universitäten und industriellen Partnern. Mit der vorliegenden Reihe haben wir versucht, ein Forum zu schaffen, das neben unseren Publikationen in Fachzeitschriften die aktuellen Forschungsergebnisse des Instituts für Thermische Strömungsmaschinen am Karlsruher Institut für Technologie (KIT) einem möglichst großen Kreis von Fachkollegen aus der Wissenschaft und vor allem auch der Praxis zugänglich macht und den Wissenstransfer intensiviert und beschleunigt.

Flugtriebwerke, stationäre Gasturbinen, Turbolader und Verdichter sind im Verbund mit den zugehörigen Anlagen faszinierende Anwendungsbereiche. Es ist nur natürlich, dass die methodischen Lösungsansätze, die neuen Messtechniken, die Laboranlagen auch zur Lösung von Problemstellungen in anderen Gebieten - hier denke ich an Otto- und Dieselmotoren, elektrische Antriebe und zahlreiche weitere Anwendungen - genutzt werden. Die effiziente, umweltfreundliche und zuverlässige Umsetzung von Energie führt zu Fragen der ein- und mehrphasigen Strömung, der Verbrennung und der Schadstoffbildung, des Wärmeübergangs sowie des Verhaltens metallischer und keramischer Materialien und Verbundwerkstoffe. Sie stehen im Mittelpunkt ausgedehnter theoretischer und experimenteller Arbeiten, die im Rahmen nationaler und internationaler Forschungsprogramme in Kooperation mit Partnern aus Industrie, Universitäten und anderen Forschungseinrichtungen durchgeführt werden.

Es sollte nicht unerwähnt bleiben, dass alle Arbeiten durch enge Kooperation innerhalb des Instituts geprägt sind. Nicht ohne Grund ist der Beitrag der Werkstätten, der Technik-, der Rechner- und Verwaltungsabteilungen besonders hervorzuheben. Diplomanden und Hilfsassistenten tragen mit ihren Ideen Wesentliches bei, und natürlich ist es der stets freundschaftlich fordernde wissenschaftliche Austausch zwischen den Forschergruppen des Instituts, der zur gleichbleibend hohen Qualität der Arbeiten entscheidend beiträgt. Dabei sind wir für die Unterstützung unserer Förderer außerordentlich dankbar.

Im vorliegenden Band der Schriftenreihe befasst sich der Autor mit der Analyse des Konzepts einer Ringbrennkammer für Flugtriebwerke mit in Umfangsrichtung angestellten Brennstoffdüsen. Dieses Konzept wird unter anderem als SHC (Short Helical Combustor) bezeichnet. Wie der Name andeutet, erlaubt bzw. erfordert die Anstellung der Brenner eine größere Höhe der Brennkammer, was bei konstantem Volumen und damit unveränderter Aufenthaltszeit des Arbeitsmediums eine kürzere und damit kompaktere Bauform der Brennkammer zur Folge hat. Ein ganz wesentlicher Vorteil der in Umfangsrichtung angestellten Brenner liegt in der Interaktion mit dem vorgeschalteten Verdichter und der nachgeschalteten Turbine. Die Umlenkung in den Austritts- bzw. Eintrittsleiträdern kann geringer ausfallen. Diese geringere Umlenkung erlaubt, das Teilungsverhältnis zu erhöhen und damit die Anzahl der Leitschaufeln zu reduzieren. Neben einer Gewichtsreduktion eröffnet sich dadurch auf der Turbinenseite die Möglichkeit, bei gegebener Turbineneintrittstemperatur entweder die Kühlluft für die Leitreihe der ersten Turbinenstufe zu reduzieren oder bei unveränderter Kühlluftmenge die Turbineneintrittstemperatur zu erhöhen. Die Analyse des Konzepts erfolgt analytisch-numerisch unter Verwendung eines quelloffenen CFD-Verfahrens für vorgemischte turbulente Verbrennungsprozesse. Von besonderer Bedeutung für die SHC-Brennkammer ist das am Brennkammeraustritt vorliegende Geschwindigkeits- und Temperaturprofil. Gewünscht sind möglichst homogene Temperatur- und Geschwindigkeitsfelder, wobei ein hoher konstanter Austrittwinkel besonders günstig ist. Zur Beurteilung verschiedener geometrischer Konfigurationen verwendet der Autor den innovativen Ansatz, den Drehimpuls der Strömung im Zusammenspiel mit der durch die Druckverteilung an den Seitenwänden hervorgerufene Reaktionskraft zu interpretieren. Durch eine geeignete Konturierung der Seitenwände gelingt es ihm, die Druckverteilung so zu beeinflussen, dass sich eine optimierte Brennkammergeometrie ergibt. Das entwickelte Brennkammerkonzept wurde zwischenzeitlich erfolgreich zum Patent angemeldet.

Karlsruhe, im August 2019   Hans-Jörg Bauer

Vorwort des Authors

Die vorliegende Arbeit entstand während meiner Tätigkeit am Institut für Thermische Strömungsmaschinen (ITS) am Karlsruher Institut für Technologie (KIT). Das Forschungsprojekt zur Untersuchung von Short Helical Combustor (SHC) wurde in Zusammenarbeit mit Rolls-Royce Deutschland, Siemens und MTU Aero Engines definiert und von der Forschungsvereinigung Verbrennungskraftmaschinen (FVV) gefördert. Ich möchte die Gelegenheit nutzen, allen zu danken, die mich in den letzten Jahren unterstützt haben.

Mein Dank gilt an erster Stelle Herrn Prof. Dr. Hans-Jörg Bauer für das entgegengebrachte Vertrauen und für die exzellenten Rahmenbedingungen, die mir am ITS geboten wurden, sowie für die Übernahme des Hauptreferats.

Ebenso möchte ich mich herzlich bei Herrn Prof. Dr. Nikolaos Zarzalis für die Übernahme des Korreferats und das Interesse an dieser Arbeit bedanken.

Mein ganz besonderer Dank gilt dem Leiter der Brennkammerabteilung am ITS und meinem wissenschaftlichen Mentor Herrn Dr. Rainer Koch. Mit seinem brillanten technischen Gespür und seiner freundlichen und zugänglichen Art hat er einen essentiellen Beitrag zum Erfolg dieser Dissertation geleistet. Außerdem waren die vielschichtigen und lebhaften Gespräche zu allen denkbaren Themen ein schöner Ausgleich zum universitären Alltag. Danke Rainer!

Bei allen Kollegen und Mitarbeitern des Instituts möchte ich mich für das ausgezeichnete Arbeitsklima am ITS, für die fruchtbaren technischen Besprechungen, sowie für die gemeinsamen Freizeitaktivitäten bedanken. Insbesondere möchte ich mich bei meinen Bürokollegen Dr. Martin Schwitzke und Manuel Hildebrandt für die schöne und angenehme Zusammenarbeit bedanken. Bei Dr. Enrico Bärow und Georg Blesinger bedanke ich mich für die Fachgespräche über Drallströmungen und die Mitteilung ihrer hervorragenden experimentellen Einsicht. Weiterhin bedanke ich mich bei Felix von Plehwe für unsere innovativen technischen Gespräche sowie unserem spannenden Austausch bezüglich unserer diversen juristischen Herausforderungen.

Darüber hinaus gilt mein Dank allen Studenten, insbesondere Sylvia Wilhelm, Simon Trapp, Marc Sebastian Schneider, Simon Holz und Felicitas Schäfer, die mit ihrem Engagement im Rahmen von HiWi-Tätigkeiten sowie Studien- und Masterarbeiten diese Arbeit bereichert haben.

Ebenso möchte ich mich bei Herrn Dr. Dimitrie Negulescu (Rolls-Royce Deutschland Ltd & Co KG) für die Initialisierung des SHC Forschungsprojekts und die gute Zusammenarbeit innerhalb der Arbeitsgruppe bedanken. Mein besonderer Dank gilt Herrn Dr. Wolfgang Meier (Institut für Verbrennungstechnik-DLR Stuttgart) für die Zurverfügungstellung der experimentellen Messdaten der Modelbrennkammer PRECCINSTA. Weiterhin geht mein herzlicher Dank an Herrn Dr. Christian Beck (Siemens AG, ITS Alumnus), der Betreuer meiner Diplomarbeit bei Siemens AG in Mülheim an der Ruhr. Von ihm habe ich viel über Gasturbinenbrennkammern gelernt und wurde zu einer Promotion am ITS motiviert.

All meinen Freunden, die durch verschiedene gemeinsame Aktivitäten die Karlsruher Zeit noch schöner gemacht haben, danke ich sehr. Dabei möchte ich namentlich Dr. Oliver Sander erwähnen, für seine Hilfe beim technischen Lektorat einiger meiner wissenschaftlichen Publikationen, aber vor allem für sein Interesse sowie seine Hilfsbereitschaft und Mitwirkung in verschieden kulturellen Veranstaltungen, die ich am KIT organisiert hatte. Weiterhin erwähne ich dankend Hajar Kamali, Dr. Saman Kiamehr, Dr. Mahtab Niknahad, Dr. Taleieh Rajabi, Hoopad Rostami, Hamid Salehifar, Peyman Toreini und Saeed Vakilzadeh, die während meiner Zeit in Karlsruhe viele gute Momente mit mir gefeiert und mich in schwierigen Zeiten unterstützt haben.

Von ganzem Herzen danke ich meinen Eltern, Manijeh Jasbi und Kiomars Ariatabar, die mir das Studium in Deutschland erst ermöglicht und mir stets den Rücken gestärkt haben.

Schließlich gilt mein herzlicher Dank meiner geliebten Ehefrau Dr. Parisa Ariatabar für ihre ununterbrochene Unterstützung und das Zurückstellen eigener Interessen während der Anfertigung dieser Dissertation.

Meschede, im August 2019   Behdad Ariatabar

To Parisa

Kurzfassung

Um die gesetzlichen Anforderungen für die Anwendung als Antrieb oder in der Energieumwand-
lung zu erfüllen müssen moderne Verbrennungssysteme eine effiziente und stabile Funktion
bei geringen Schadstoffemissionen über einen weiten Betriebsbereich gewährleisten. Neuartige
Brennkammerkonzepte, die grundlegenden aerothermischen und strukturmechanischen Vorteile
aufweisen können, sind daher in der Energietechnik von besonderem Interesse.

In der vorliegenden Forschungsarbeit wird ein innovatives Ringbrennkammerkonzept für Flugtriebwerke untersucht. Das Hauptmerkmal dieser Brennkammer ist die staffelförmige Anordnung
der Brenner, die ein spiralförmiges Strömungsprofil im Flammrohr impliziert. Eine solche
Strömungsstruktur mit hohem Austrittsströmungswinkel kann angepasst werden, um die axiale
Länge der Brennkammer sowie die Anzahl der vor- und nachgelagerten Schaufeln zu reduzieren.
Dadurch können die Gesamtleistung sowie der Wirkungsgrad eines solchen Triebwerks aufgrund
mehrerer aerothermischen und strukturmechanischen Vorteile verbessert werden.

Das Primärziel dieser Forschungsarbeit ist die Analyse der reagierenden spiralförmigen Strömung
in einer solchen Ringbrennkammer mit gestaffelter Brenneranordnung anhand numerischer Meth-
oden. Darüber hinaus wird die Empfindlichkeit der Strömungsstruktur hinsichtlich ausgewählter
relevanter Designparameter untersucht. Basierend auf den gewonnenen physikalischen Erken-
ntnissen wird eine neuartige Methode zur Strömungsoptimierung vorgestellt und geprüft. Ziel
dieser Methode ist die Führung der Strömung basierend auf der Manipulation des Druckfeldes in
der primären Verbrennungszone durch Konturierung der Linerwände.

In einem ersten Schritt werden basierend auf ähnlichkeitsbetrachtungen geeignete geometrische
Konfigurationen sowie aerothermische Randbedingungen anhand einer Parameterstudie fest-
gelegt. Alle Konfigurationen liefern am Austritt jedoch eine ungleichförmige Strömung mit
sehr niedrigem gemittelten Strömungswinkel. Die Auswertung des integralen Drallsatzes zeigt,
dass der enorme Rückgang des Strömungswinkels in der Austrittsebene der Brennkammermod-
elle in erster Linie auf ein hohes Reaktionsdrehmoment zurückzuführen ist, welches durch die
asymmetrische Verteilung des Staudrucks an der gestaffelten Brennkammerwand verursacht
wird.

Im zweiten Schritt wird ein Brennkammermodell mit einer generisch konturierten Wand entwick-
elt und simuliert, um das Potenzial der Strömungsführung durch Manipulation des Druckfeldes
zu untersuchen. Das Brennkammermodel mit konturierter Linerwand zeigt im Vergleich zum
Referenzbrennkammermodell einen um ca. 50% höheren gemittelten Ausströmungswinkel mit
wesentlich gleichförmigerer Strömung in seiner Austrittsebene.

Die in dieser Arbeit demonstrierte Methodik zur Brennkammerströmungsführung ist ein essen-
tieller Schritt zur Erreichung der technischen Reife für die Anwendung des vorgeschlagenen
Brennkammerkonzepts in einem Flugtriebwerk. Darüber hinaus können mit dem gleichen Ansatz
Form, Größe und die Lage der Flammen auch in anderen Verbrennungssysteme kontrolliert
werden. So ist auch bei den konventionellen Gasturbinenbrennkammern eine Verbesserung der
Verbrennungsstabilität, der Schadstoffemissionen und der Brennkammer-Turbinen-Interaktion
zu erwarten.

Abstract

Modern combustion systems must ensure an efficient and stable operation with low emission of air pollutants over a wide range of conditions to meet the regulatory requirements for their application in propulsion and power generation. Novel combustor concepts, which can provide major aerothermal and structural advantages are, therefore, of particular interest in energy technology.

In the present study, an innovative annular aeroengine combustor concept is investigated. The main feature of this combustor is the staggered arrangement of the burners, which implies a helical flow pattern in the flame tube. Such a flow structure with high exit flow angle can be adopted to reduce the axial length of the combustor as well as the number of the up- and downstream vanes. In consequence, the overall performance and efficiency of such an engine can be improved due to several aerothermal and structural advantages.

The primary goal of this research study is to investigate the helical reacting flow in such an annular combustor with a staggered arrangement of the burners via numerical methods. Moreover, the sensitivity of the flow structure to some relevant aerothermal and geometry design parameters will be studied. Based on the physical insight gained, a novel flow control method is proposed and investigated. This method aims at the treatment of the combustor flow based on the manipulation of the pressure field in the primary combustion zone by contouring the liner walls.

In the first step and based on similarity considerations, the geometric scaling and aerothermal boundary conditions are determined via a parametric study. However, the flow at the exit of all configurations exhibited high aerothermal nonuniformities and low average flow angles. The underlying physics of the complex helical flow is elucidated with the integral analysis of the balance equation for the angular momentum. It is shown that the tremendous decrease of the average flow angle at the combustor exit plane is primarily attributed to a high reaction torque caused by the asymmetric distribution of the stagnation pressure at the staggered combustor wall.

As the next step, a combustor model with a generic contoured wall is developed and simulated to investigate the potential for flow control via manipulation of the pressure field at the primary combustion zone. Compared with the reference combustor model, the contoured liner wall model exhibited an approximately 50% higher average outlet flow angle with substantially better aerothermal uniformity at the exit plane.

The demonstrated flow control method provides an important step toward attaining the technical maturity to employ the investigated novel combustor concept in an aircraft engine. Moreover, with the same approach, the shape and the size, as well as the location of the flames, can also be manipulated in other combustion devices. Accordingly, an improvement of the combustion stability, emission characteristics, and the combustor–turbine interaction is expected for the conventional gas turbine combustors as well.

Contents

Nomenclature

Acronyms

ATF	Artificially Thickened Flame
CFD	Computational Fluid Dynamics
CFL	Courant-Friedrichs-Lewy Number
CPU	Central Processing Unit
CPUh	CPU hour
CRTS	Constant Residence Time Scaling
CSW	Contoured Sidewall
CV	Control Volume
CVS	Constant Velocity Scaling
DA	Double Annular
DM	Dome
FSD	Flame Surface Density
FSW	Flat Sidewall
HPC	High Performance Computing
IL	Inner Liner Wall
IRZ	Inner Recirculation Zone
KIT	Karlsruhe Institute of Technology
LDV	Laser Doppler Velocimetry
LES	Large Eddy Simulation
LHS	Left Hand Side
NGV	Nozzle Guide Vanes
OGV	Outlet Guide Vanes
OL	Outer Liner Wall
ORZ	Outer Recirculation Zone
PaSR	Partially Stirred Reactor
PDF	Probability Density Function
PVC	Precessing Vortex Core
RANS	Reynolds Averaged Navier–Stokes
RCC	Reference Conventional Combustor
RHS	Right Hand Side
RMS	Root Mean Square
RNG	Renormalization Group
RSM	Reynolds Stress Model
SA	Single Annular
SGS	Sub Grid Scale
SHC	Short Helical Combustor
SST	Shear Stress Transport
SV	Singular Values
SW	Sidewall
TFC	Turbulent Flame Speed Closure

TSM		Two Step Mechanism

Symbols

A	m^2	Surface area
a		General vector or tensor property
α	$^\circ$	Flow angle against the engine rotor axis
β	$^\circ$	Burner tilting angle against the engine rotor axis
C_h	$-$	Enthalpy factor (Eq.(4.18))
C_l	$-$	Linear momentum factor (Eq.(4.22))
C_{lm}	$-$	Linear momentum/mass flow factor (Eq.(4.24))
C_L	$-$	Angular momentum factor (Eq.(4.23))
C_{Lm}	$-$	Angular momentum/mass flow factor (Eq.(4.26))
C_m	$-$	Mass flow factor (Eq.(4.21))
c	$-$	Reaction progress variable
c_ε	$-$	Constant in the model equation for ε (Eq.(A.11))
$c_{\varepsilon 1}$	$-$	Constant in the model equation for ε (Eq.(A.11))
c_{kpp}	$-$	KPP mean reaction rate model constant (Eq.(2.35))
c_μ	$-$	Turbulent viscosity constant in k,ε-model (Eq.(A.9))
c_p	$J/kg/K$	Specific heat capacity at constant pressure
c_{st}	$-$	Turbulent flame speed model constant (Eq.(2.34))
Da	$-$	Damköhler number
d	m	Burner characteristic diameter
dA	m^2	Surface element vector pointing along the outward normal
dV	m^3	Volume element
Δ	$-$	Flame tube/burner confinement ratio
$\mathcal{D}_e$	m^2/s	Effective diffusivity
$\mathcal{D}_f$	m^2/s	Flame diffusivity
$\mathcal{D}_m$	m^2/s	Mass (molecular) diffusivity
$\mathcal{D}_T$	m^2/s	Thermal diffusivity
$\mathcal{D}_t$	m^2/s	Turbulent mass diffusivity
ε	m^2/s^3	Dissipation rate of turbulent kinetic energy
F_{ext}	N	Sum of external forces
f_v	N/m^3	External volume forces vector
Φ_A		Surface source of the scalar ϕ
Φ_V		Volume source of the scalar ϕ
ϕ		General scalar property
g	m/s^2	Gravitational acceleration vector
γ	$-$	Isentropic exponent
H	m	Height of flame tube segment
H_t	J	Total enthalpy
h_t	J/kg	Specific total enthalpy
I	N	Linear momentum flow
I		Unit tensor

Symbol	Units	Description
$\mathcal{J}_\phi$		Diffusive flux of scalar ϕ
Ka	$-$	Karlovitz number
k	m^2/s^2	Turbulent kinetic energy
L	$N \cdot m$	Angular momentum flow
L	m	Length of flame tube segment
Le	$-$	Lewis number
L_p	$N \cdot m$	Pressure torque
L_τ	$N \cdot m$	Friction torque
ℓ	m	Turbulent length scale
ℓ_δ	m	Laminar flame (chemical) reaction layer length scale
ℓ_f	m	Laminar flame (chemical) length scale
$\ell_\mathcal{I}$	m	Integral length scale
ℓ_κ	m	Kolmogorov length scale
$\mathcal{L}$	m	Geometric length scale
λ	$W/m/K$	Heat conductivity
M	$-$	Mach number
m	kg	Mass
$\dot{m}$	kg/s	Mass flow
μ	$kg/m/s$	Dynamic viscosity
μ_t	$kg/m/s$	Dynamic eddy viscosity
N_b	$-$	Number of burners
N_{seg}	$-$	Number of flame tube segments
ν	m^2/s	Kinematic viscosity (viscous momentum diffusivity)
ν_t	m^2/s	Kinematic eddy viscosity (turbulent momentum diffusivity)
$\mathcal{O}$		Order of magnitude
ω_c	$kg/m^3/s$	Reaction rate
P	m	Pitch of flame tube segment
P_s	Pa	Static pressure
P_t	Pa	Total pressure
π	$-$	Ratio of flame tube length to the axial extent of the recirculation bubble
Q_V	J/m^3	Volume energy sources
q	J/m^2	Heat flux vector
R	m	Radius
Re	$-$	Reynolds number
r	m	Position vector
ρ	kg/m^3	Density
S	$-$	Swirl number
s_L	m/s	Laminar burning velocity
s_T	m/s	Turbulent burning velocity
σ	N/m^2	Stress tensor
σ_α	$^\circ$	Standard deviation of the flow angle α
σ_ε	$-$	Constant in the model equation for ε (Eq.(A.11))
σ_k	$-$	Constant in the model equation for ε (Eq.(A.11))

T	K	Temperature
T_b	K	Adiabatic flame temperature
T_0	K	Temperature at inner flame layer
T_t	K	Total temperature
t	s	Time
t_f	s	Laminar flame (chemical) time scale
$t_{\mathcal{I}}$	s	Integral time scale
t_κ	s	Kolmogorov time scale
τ	N/m^2	Viscous shear stress tensor for Newtonian fluid
U	m/s	Time-averaged velocity component
$\boldsymbol{u}$	m/s	Velocity vector
V	m^3	Volume
$\dot{V}$	m^3/s	Volumetric flow
$\dot{W}$	W	Turbine power output
x_{rb}	m	Axial extent of recirculation bubble
Y	$-$	Species mass fraction
r, θ, z	m, rad, m	Cylindric coordinates
x, y, z	m, m, m	Cartesian coordinates
$\partial f/\partial x$		Partial derivative
$f \sim g$		f scales with g
∇		Nabla differential operator
$\boldsymbol{uv}$		Outer product of matrices $\boldsymbol{u}$ and $\boldsymbol{v}$
$\cdot$		Inner product
$:$		Double inner product

Superscripts

$\bar{q}$		Averaged quantity
$\tilde{q}$		Favre-averaged quantity
q'		Fluctuation around the averaged value
q''		Fluctuation around the Favre-averaged value

Subscripts

q_b		Burned flow condition
q_β		Quantity related to tilting angle β
q_i		Quantity at an inlet station
q_{in}		Inflow quantity
q_o		Quantity at an outlet station
q_{out}		Outflow quantity
q_u		Unburned flow condition

1   Introduction

1.1   Motivation

Global air traffic has doubled in size every fifteen years in the past four decades and will continue to do so (ICAO, 2016). Gas turbines are the primary power source for aircraft propulsion (Farokhi, 2014). Furthermore, the application of power generating gas turbines may grow due to the increasing use of renewable energies. In that scenario, fast-starting gas turbines with highly efficient part-load capability can provide a remedy to cover the peak times or situations with a temporary lack of renewable sources such as wind or sun.

The increasing application of gas turbines for propulsion and power generation faces permanently stricter regulations for pollutant emissions (BDL, 2017, EPA, 2017). To secure profitable operation, as well as to meet the regulatory requirements, gas turbine manufacturers have to improve low-emission combustion technologies. Besides, the overall efficiency of the engines must be continuously increased to reduce fuel consumption and associated emissions. In both of these efforts, the combustor plays a central role and is the subject of numerous research and development activities. In addition to these, aviation pursues the reduction of its environmental impacts by improving the aerodynamics of airplanes as well as by developing more efficient routes facilitated by performance-based procedures and advanced avionics.

The overall efficiency of gas turbines can be increased by:

1. Improving the thermal efficiency of the simple Joule–Brayton cycle by higher pressure ratios and higher turbine inlet temperatures.

2. Carnotization of the simple Joule–Brayton cycle: Recuperation of the exhaust gases, multistage compression with intercooling, multistage expansion with sequential combustion.

3. Enhancing the efficiency of the engine components.

4. Optimizing the interaction of the engine components.

5. Raising the propulsive efficiency, in the case of aircraft engines, by a higher bypass ratio.

The gas turbine combustor interacts directly with two other major engine components, i.e., the compressor and the turbine. During the progress of the flow through the combustor, the working fluid experiences aerothermal changes with velocity and temperature variations spanning over an order of magnitude. The combustion device must ensure an efficient and safe functionality at different operating points, which can be far apart from each other. Such a combustor requires an optimal treatment of several multiphysics phenomena involving complex flow guidance, preparation and mixing of fuel with air, cooling of liner walls, and expressly low-emission and stable combustion.

As a central component of the gas turbine, the combustor is involved in numerous innovative concepts. The subject of the present research work is a compact combustor with angular air supply. This concept adopts measures to improve the interaction of the combustion system with the upstream compressor and downstream turbine. In the next section, the technical history of this idea is reviewed, and its principal features are introduced.

1.2   State of the Art in Gas Turbine Combustion

Traditionally, the gas turbine combustion systems used nonpremixed flames because of their reliable performance and stability characteristics. The severe drawback of this type of combustor is the production of very high levels of thermal nitrogen oxides (NOx) (Correa, 1998). Modern machines employ low-emission combustion technologies, including lean-burn (LB) and rich-burn quick-quench lean-burn (RQL). RQL techniques are applied widely in propulsive engines because of a good compromise between the emission and stability characteristics. They are mainly hampered by soot formation and incomplete mixing between fuel-rich combustion products and air. Thus, the RQL technology faces increasing challenges to comply with the current emission regulations and seems to reach its technological limit in the foreseeable future. Despite promising emission characteristics, the catalytic combustion could not gain acceptance because of the high cost as well as durability and safety issues (Lefebvre and Ballal, 2010).

Among these three technologies, LB combustion appears to be the most promising technology for most practical systems at present. It is employed in nearly all land-based and increasingly in large aircraft engines but also in boilers, furnaces, and internal combustion engines. LB technology features first and foremost reduced NOx emissions. The reason is the operation of the primary combustion zone with excess air. Accordingly, the average flame temperature, and therefore the thermal NOx formation, is reduced. (Dunn-Rankin, 2011).

The improved emission properties of the LB concept are at the costs of low reaction rates, the hazard of lean blow out, high sensitivity to mixing quality as well as thermoacoustic instabilities. The latter are unsteady flow oscillations, which can reach sufficient amplitudes to interfere with engine operation. In extreme cases, they cause total failure of the system due to excessive structural vibration and heat transfer to the chamber (Huang and Yang, 2009). These issues can seriously impair the stability and reliability of the LB combustion. Thus, in contrast to land-based gas turbines, the widespread application of LB combustion in aircraft engines still requires more maturing of this technology. Nevertheless, with respect to a long-term application, the unconventional combustor model addressed in the present thesis is studied in the context of LB technology.

1.2.1   Short Helical Combustor

The subject of the present research work is the concept of an annular gas turbine combustor as shown schematically in Fig. 1.1. The central idea is the utilization of angular momentum of the airflow discharging from the rotor blades of the last compressor stage. Accordingly, the Outlet Guide Vanes (OGV) do not deflect the flow entirely in the axial direction, as is done in the case of conventional machines. Thus, the flow discharges into the combustor with a circumferential

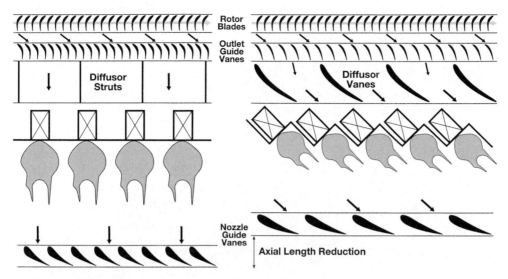

Figure 1.1: Schematic comparison of a conventional annular gas turbine combustor (left) with the Short Helical Combustor (right). Solid arrows indicate the flow path. Reproduced from Negulescu (2013).

component. In the case of an annular combustor, a helical flow pattern would be established. This property, in turn, can be adopted to reduce the axial length of the combustor.

The overall performance and efficiency of such an engine can be improved by several aerothermal and structural properties. In particular, because of the shorter axial extent of the combustor, the total length of the engine can be reduced. In addition to the weight reduction, better integration of the combustor and a more rigid shaft with improved rotor dynamics are expected as well. Furthermore, because of the tilted burner arrangement, a smaller deflection angle is required at the vane cascades upstream and downstream of the combustor. Accordingly, the number of OGV and Nozzle Guide Vanes (NGV) can be reduced. Thereby, smaller total pressure loss, as well as lower cooling air demand is expected. Another substantial advantage is the enhanced transversal exchange of heat and combustion products between the adjacent flames via advection and radiation. Hence, a reduction of pollutant and noise emissions together with higher stability and startup flexibility might be achieved.

Seippel (1943) was the first to patent a gas turbine combustor design with such an angular air supply. In a similar context, Negulescu (2013) proposed the Short Helical Combustor (SHC). Many other similar configurations have also been patented by almost all major gas turbine manufacturers (Burd and Cheung, 2007, Buret et al., 2009, Hall, 1961, Mancini et al., 2007, Schutz et al., 1999, Tangirala and Joshi, 2015). Commonly, in all those publications, the well-established concept of swirl-stabilized combustion was followed. Accordingly, the combustor dome features a staggered arrangement. The axes of burners were then tilted against the rotor axis to adapt to the angled flow coming from the upstream compressor. A new element due to the staggered design is the sidewall. It is located in the primary combustion zone and gives rise

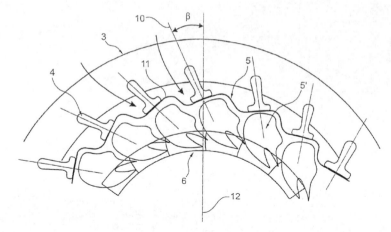

Figure 1.2: The SHC concept for a centripetal combustor (Negulescu, 2014)

to several issues regarding heat transfer, aerodynamics, as well as static and dynamic combustion stabilities.

The concept of compact combustors with integrated compressor outlet and turbine inlet has also been followed in the context of the so-called Trapped Vortex or Ultra-Compact Combustor by Bohan and Polanka (2013), Burrus et al. (2001), Hsu et al. (1995, 1998), Roquemore et al. (2001), Strokin et al. (2013, 2017), Zelina et al. (2002, 2004). The main advantage of this concept is the elimination of sidewalls. However, such combustion devices particularly face stability issues at high thermal loads, as well as efficiency and emission characteristics, which presently do not meet the regulations for modern civil aircraft engines or industrial and power plant gas turbines.

The operation of a conventional annular combustor would be substantially affected by its adaptation to the SHC design. Various new multiphysical phenomena with counteracting effects might occur. The primary design issues to be considered are

- **Diffuser:** The airflow from the low-turn OGV has a circumferential velocity component in the diffuser, where its axial part becomes smaller due to the deceleration. Thus, the mean flow angle becomes larger. As shown in Fig. 1.1, the diffuser struts need to be adapted to the three-dimensional (3D) flow structure to minimize the aerodynamic losses as well as to provide a uniformly distributed airflow at the inlet of the swirl burners. The latter is vital for stable operation and optimal emission characteristics of the adjacent burners.

- **Heat transfer:** The volume to surface ratio of the combustor is a fundamental parameter regarding its heat transfer characteristics. Comparing with a conventional combustor, an equivalent SHC with the same volume could have a smaller total surface area, and thus a lower overall wall heat load. However, the local heat transfer to the corners of the staggered wall can be extremely high and may be critical.

- **Combustion:** The interaction of the flames and the sidewalls may result in flame quenching with negative implications for combustion stability and emissions. On the other hand,

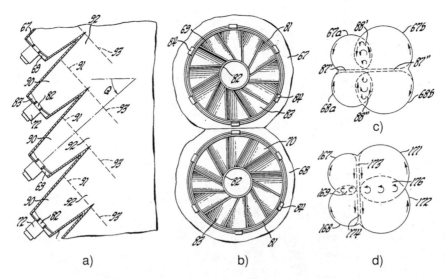

a) b) d)

Figure 1.3: a) Double annular combustor with angular air supply. b) Radially counterrotating adjacent swirlers. c,d) Flow analysis on the cross section denoted by the number 92. Radially counterrotating and circumferentially corotating swirls as depicted in c) result in a convenient circumferential flow at the mid-height of the flame tube (87), and flame stabilizing by mixing of the hot gases of the adjacent burner (88).

because of the helical arrangement, each flame root will be heated by advection as well as by radiation and supplied with reactive species from the adjacent burner. Similar to exhaust gas recirculation, an improvement of the blowout and stability limits is expected. This *piloting* effect is in particular advantageous for lean combustion. In addition to improved lean blowout stability, a better lightup and relight capability, as well as emission characteristic for the SHC, might be achieved.

The realization of the SHC requires an elaborate analysis and quantification of all these effects and their interactions. The existing patents are focused on schematic illustrations of the SHC and general descriptions of its aerothermal advantages. In this context, different possibilities to install the burners have been proposed by Burd and Cheung (2007). Negulescu (2013) performed an aerodynamical analysis of the SHC with focus on the inlet to the combustor, i.e., OGV and diffuser. In a subsequent study, Negulescu (2014) proposed an integration of the SHC concept in a centripetal combustor as shown in Fig. 1.2. This is particularly convenient because the conservation of angular momentum implies a significant rise of the circumferential flow velocity by streaming from higher to lower radii. Furthermore, the arrangement of the burners on the larger radius offers more place for higher tilting of the burner axes. Buret et al. (2009) focused on minimizing the aerodynamic losses and proposed inclined cooling orifices on the combustor liner. Accordingly, the angles of inclination were matched to the flow angle within the inner and outer annulus of the SHC. Hall (1961) proposed a double annular burner arrangement as shown in Fig. 1.3. He studied different configurations of the adjacent burners with the goal to optimize the mixing in the primary combustion zone. He suggested, 53 years before Ariatabar et al. (2014), a counterrotating swirl configuration of radially adjacent burners based on brilliant aerodynamic deliberations. However, these patents do not address what maximum tilting angle can be realized in an SHC and what the effects on the flow pattern and combustion characteristics are.

The study presented here is a detailed analysis and assessment of the SHC concept concerning its capability of reducing the combustor length as well as the number of the NGV. The focus will be on the aerothermal characteristics of the swirl flames because they might be strongly affected by the staggered combustor embodiment. It is known that the intense swirling flows are extremely sensitive to the external boundary conditions (Delery, 1994, Leibovich, 1978, 1984). Thus, the asymmetric confinement of the swirling flames, as occurs in the SHC due to the sidewalls, might fundamentally alter the structure of the flow and so impair the combustion characteristics.

In the following section, the effects of inflow and geometric boundary conditions on the structures of nonreacting and reacting swirling flow fields are briefly reviewed.

1.2.2   Asymmetrically Bounded Swirling Flows

Swirl-stabilized combustion is widely employed in gas turbines. The reason is the high reaction rate and combustion stability accompanied by excellent burnout and emission characteristics. As shown in Fig. 1.4, swirling flows exhibit a hydrodynamic instability in the form of an inner recirculation zone (IRZ) under certain conditions. The occurrence of this flow phenomenon is called "vortex breakdown," which acts in the case of combustion as an aerodynamic flame-holder. The corresponding flow reversal is responsible for the superior stability and efficiency of swirling flames. Accordingly, the upcoming fresh mixture is continuously ignited by the recirculating hot gases, while the shear layers between the swirling jets and the recirculating flow ensure an intensive turbulent mixing. Both the formation of the flow reversal, as well as the intensity of turbulent shear layers, depend upon the swirl strength. Therefore, usually medium- to high-swirl burners are used in gas turbine combustors. By an increased swirl strength, the coupling between velocity components and pressure field becomes stronger and more complex. Hence, the flow structure becomes more sensitive to inlet, boundary, and geometry effects (Lucca-Negro and O'doherty, 2001).

The structure of the swirling flame is an important aspect, which must be considered in the design of a gas turbine combustor. Suboptimal flame structures can impair both the static and dynamic stability of a combustor in the form of blowoff and combustion instabilities, respectively. A vast body of experimental and theoretical studies address single swirling flames with symmetric boundary conditions. On that basis, supported by operational experiences, the gas turbine manufacturers often follow their own design principles to control the structure and stability of swirling flames. Thereby, most of the operating gas turbines for propulsion and power generating applications use robust combustor configurations with axisymmetric inflow and geometric boundaries for the swirl burners.

On the other hand, the application of innovative combustor designs in modern gas turbines is

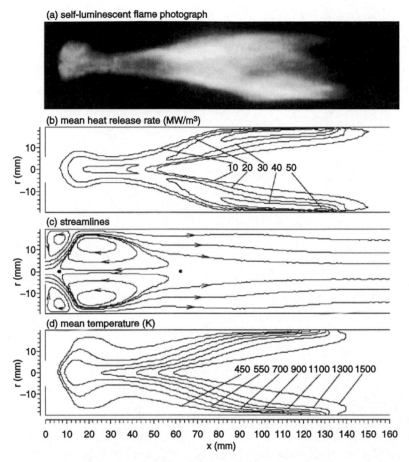

Figure 1.4: Computed contours and self-luminescent photograph for an equivalence ratio of 0.59. a) Self-luminescent flame photograph by a downstream ignition, b) contours of mean volumetric heat release rates, c) streamlines, d) isotherms. Mean axial entry velocity of 10 m/s, Swirl number of 0.72. (Bradley et al., 1998)

often associated with a complex arrangement of burners. These configurations lead inevitably to asymmetric geometric confinement as well as aerothermal boundary conditions for each swirling flame. Such a prominent example is the double annular combustor as schematically illustrated in Fig. 1.5. Thereby, the inner and outer swirling flames are subjected to asymmetric thermal and aerodynamic boundaries. The former is a consequence of liner cooling, whereas the latter is due to the different rotational direction and swirl intensity of radially adjacent burners. The whole situation of asymmetrically bounded swirling flows is schematically highlighted by the profile of the resulting azimuthal velocity (V_θ), colored by temperature, plotted over the radial direction. In the specific case of the SHC, the asymmetric confinement of the swirl flames due to the sidewalls might jointly bring about all of these effects.

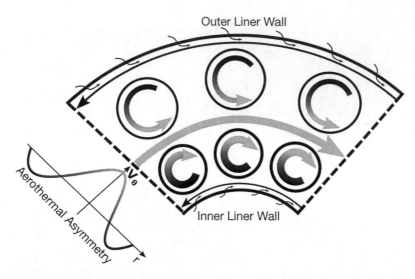

Figure 1.5: Schematic illustration of a double annular combustor. V_θ denotes azimuthal velocity. Arrows indicate the direction, and their thickness the strength, of the velocity vector. Arrows through the walls represent the liner cooling air. The dark color refers to lower and light to higher temperatures. Dashed lines are periodic boundaries.

In contrast to symmetrically bounded swirling flows, there are only a few research studies explicitly addressing the asymmetric effects. For the review in this section, to begin, some representative papers are selected. They point out the sensitivity of isothermal and reacting swirling flows to the inflow and geometric boundary conditions. Afterward, some recent experimental investigations are introduced. Thereby, the effect of asymmetric boundaries on the flow structure can be qualitatively deduced from the flow field of multiswirler arrangements.

More than six decades of intensive research on swirling flows has resulted in numerous publications. However, it is remarkable that basic physical mechanisms underlying the vortex breakdown are not yet commonly accepted. The descriptions of the governing effects, and proposed criteria to predict the onset and structure of the flow reversal, vary considerably from one publication to another. Among others, Escudier and Zehnder (1982) tried to propose a *simple* criterion for the occurrence of vortex breakdown. Their approach was based on the theoretical arguments of Benjamin (1962, 1967) and supported by similarity concepts and experimental observations. Their specific test rig featured a swirling water jet discharged into a large tank. They claimed that the onset of flow reversal and the corresponding vortex regimes could be predicted with a single parameter, which exclusively involved the circulation number, the ratio of radial to tangential velocities in the inflow region, and the Reynolds number. Later, Billant et al. (1998) also investigated swirling water jets following similar theoretical arguments as Escudier and Zehnder (1982). However, for the same parameter values and the same nozzles, they observed new vortex phenomena occurring randomly from one experiment to another and being extremely sensitive to temperature variations in the inflow. The investigations of many other researchers including the later publications of Escudier and co-workers, e.g., Escudier (1988), confirmed that

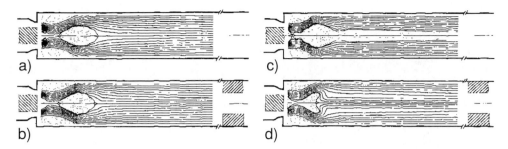

Figure 1.6: Influence of an exit contraction on swirling flow in a combustor geometry: a) supercritical flow with no exit contraction b) supercritical flow with exit contraction (54.5% diameter reduction) c) subcritical flow with no exit contraction d) subcritical flow with exit reduction (Escudier and Keller, 1985).

the onset and structure of vortex breakdown depend decisively on many other parameters. The most important ones are the inflow profile of the tangential velocity, the confinement properties, and the swirl strength. The latter can be represented by the swirl number as proposed by Beér and Chigier (1972)

$$S = \frac{2G_\phi}{RG_x} \tag{1.1}$$

where G_ϕ stands for the axial flux of angular momentum, G_x for the axial flux of the linear momentum, and R for a characteristic radius of the swirl nozzle.

In this context, Escudier and Keller (1985) investigated the effect of contraction of the downstream flow cross section on the structure of the recirculation zone as shown in Fig. 1.6. They found that for subcritical flow, the information of the downstream geometry can be propagated upstream and affect the vortex breakdown, somewhat like acoustic waves in subsonic flow. Thereby, as a rule of thumb established by Squire (1960), the flow is subcritical if the maximum tangential velocity exceeds the axial velocity. For a more general assessment, it is necessary to use the critical equation given by Benjamin (1962).

Farokhi et al. (1989) experimentally proved that the swirl number as an integral parameter is insufficient in describing the character of the swirling flow. They showed that the initial tangential velocity profile, as a distribution between the free- and forced-vortex types, has a significant effect on the evolution of confined swirling jets. Furthermore, the nozzle/chamber diameter ratio in a sudden expansion was identified as a deciding parameter influencing the onset of vortex breakdown and the downstream flow structure in the measurements of Hallett and Toews (1987). Chao et al. (1991) showed a significant spectral and structural sensitivity of swirling flow field to nonaxisymmetric disturbances. They found that the downstream azimuthal instability promotes the vortex breakdown. Cary et al. (1998) numerically examined the influence of weak asymmetric disturbances on an axisymmetric swirling flow. They tracked the evolution of azimuthal asymmetries in the Fourier-decomposed incompressible Navier–Stokes equations. Beginning with an axisymmetric solution, they added an asymmetric disturbance at the inlet of a converging–diverging nozzle and allowed it to propagate. For axisymmetric swirling flows without breakdown, they found that the perturbations were merely convected and dissipated. The presence of vortex breakdown in the case of larger swirl numbers, however, amplified the asymmetric disturbances and led to massive changes in the flow structure and morphological alteration of the recirculation bubbles.

From the above-reviewed publications, with the focus on nonreacting flows, it can be concluded that the structure and main feature of the swirling flows are highly sensitive to the imposed boundary and initial conditions. Moreover, it is known that the heat release due to combustion can also fundamentally alter the structure of the swirling flow field (Rusak et al., 2002, Stein and Kempf, 2007). The density of the combustive gases decreases due to the temperature increase. This leads to gas expansion, in general, alongside the pressure gradient toward the combustor outlet. The average axial flow velocity is increased. The average azimuthal velocity might decrease in compliance with the balance equation for angular momentum. It would take place as a consequence of the radial expansion of a swirling jet. Hence, overall, with neglecting the slight diffusive changes, combustion leads to an increase in linear momentum flow whereas angular momentum flow is almost unchanged. As a result, the swirl number is reduced. Thus, a major criterion for the formation of vortex breakdown is diminished. On the other hand, the analysis of vortex dynamics in swirl-stabilized combustion revealed additional mechanisms that promote flow instability. The so-called "combustion-induced vortex breakdown" occurs due to the negative gradient of the azimuthal vorticity in the flow caused by the flame/vortex interaction (Burmberger and Sattelmayer, 2011, Fritz et al., 2004, Kröner et al., 2007).

As such, both the time-averaged shape of the IRZ and the transient features of the flow are affected by the heat release. It can be generally concluded that coupling between the flow and flame dynamics occurs through several mechanisms. It includes wobble and precession of the flow and flame coupled with variations in the size and shape of the IRZ arising from changes in the swirl number. The configuration and operating conditions of the combustor, however, also play a decisive role, so that universal comments on the effect of combustion on swirling flows are often limited to generic setups. In this respect, Renard et al. (2000), Syred (2006), Syred and Beer (1974) provided comprehensive reviews concerning swirl-stabilized combustion.

Multiarray swirler configurations can provide a suitable framework to study the effect of asymmetric boundaries on swirling flames. However, relatively few studies have been conducted with a focus on such arrangements. Most of the related publications do not address the physics underlying the specific characteristics of swirling flows in such cases. The focus instead was predominantly on the phenomenological description of the flow fields. In addition, the combustion efficiency and stability, as well as emission characteristics of different swirl configurations, were assessed via global evaluations.

In that regard, Cai et al. (2002) measured the mean velocity and turbulent kinetic energy characteristics of a 3×3 swirler matrix. They studied co- and counterrotating configurations using LDV. As shown in Fig. 1.7, the corotating case exhibits a considerable deformation of the recirculation zones, already one swirler diameter downstream of the inlet. A further three swirler diameters downstream, the unstable vortex structures collapsed and merged into a large rotating flow structure. In contrast, the swirling flows in the counterrotating arrays retained their char-

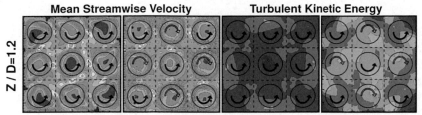

Figure 1.7: Laser Doppler velocimetry (LDV) measurements of nonreacting flow: Contours of mean streamwise velocity and turbulent kinetic energy on a cross-sectional plane downstream at Z/D = 1.2 of a 3×3 swirler matrix in co- and counterrotating arrangements (Cai et al., 2002). D refers to the diameter of the swirlers, Z to the streamwise direction. The LDV plots are based on gradient color scales, blue refers to minimum and red to maximum values.

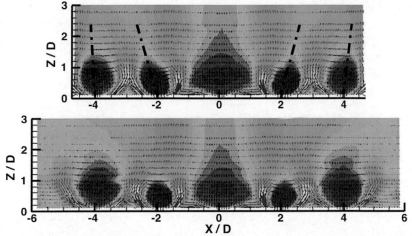

Figure 1.8: LDV measurements of a nonreacting flow: Mean streamwise velocity contours for corotating five-swirler arrangements with endwall distances of 0.75D and 2D. Dashed black lines indicate the deflection of swirl axes (Kao et al., 2014). D refers to the diameter of swirlers, Z to streamwise- and X to the radial direction. The LDV plots are based on gradient color scales, blue refers to minimum and red to maximum values.

Figure 1.9: Reacting flow: Photograph (left), 3D reconstruction (center), and cross-sectional CH* chemiluminescence image of a swirl-stabilized flame in a five-swirler combustor (Samarasinghe et al., 2013). Arrows show the rotational direction of swirling flames. The chemiluminescence plot is based on gradient color scales, black refers to minimum and red to maximum values.

acteristics across a longer distance. This stability seems to be due to the amplified azimuthal velocities at the adjoining boundaries, which is also reflected in the higher levels of turbulent kinetic energy. Thus, the counter rotating configuration featured, at far downstream distances, almost axisymmetric flow structures. Thereby, the symmetric and undeformed flow structures are expected to be advantageous for static and dynamic combustion stabilities. Moreover, the higher turbulence levels provide better mixing and might improve the emission characteristics in the case of combustion.

Kao et al. (2014) studied the aerodynamics of linearly arranged five corotating swirler arrays. They varied the interswirler spacing, as well as the endwall distance, and provided a phenomeno-logical description of the measured velocity fields. The results of LDV measurements for two cases with the same interswirler spacing and different endwall distances are shown in Fig. 1.8. It can be recognized that the tighter confinement of the corner swirlers, located at the left and right ends, also affected the adjacent inner swirling flows. The corner recirculation bubbles became compacter, and their axes were deflected toward the walls. As a counterbalance consequence, the size and shape of the adjacent recirculation bubbles were changed like the domino effect. This experimental observation mainly emphasizes the sensitivity of swirling flows for unilateral confinement, which also happens in the SHC due to the sidewall.

Finally, some of the experimental findings of Samarasinghe et al. (2013) should be introduced here. Their work was focused on the development of a tomographic image reconstruction technique. Their goal was a 3D measurement of the CH* chemiluminescence distribution in turbulent flames. Selected results are presented in Fig. 1.9, which were obtained in a lean-premixed combustion experiment. It included four peripheral and one central arranged corotating swirlers in a cylindrical model combustor. While only the central swirling flame is subject to symmetric boundary conditions, the surrounding outer flames are asymmetrically sheared by the cylindrical wall at their outer boundaries. Elsewhere on their circumference, they are sheared by adjacent swirling flows with varying intensities. The visualizations of the multiswirl combustion reveal an axisymmetric flame structure at the center. In contrast, the circumferential flame patterns are asymmetrically deformed and have significantly enlarged recirculation zones.

The review of these experimental observations should reveal the substantial effects of asymmetric boundaries on the structure of isothermal and reacting swirling flows. In the SHC, the swirling flames are unavoidably subjected to asymmetric boundaries. With this background, it is likely that the flames exhibit undesirable properties such as significant deflection and deformation of the recirculation bubbles as schematically illustrated in Fig. 1.10. This distorted flow field would lead to suboptimal combustion characteristics and nonuniform flow and temperature patterns at the exit of the combustor.

For the industrial realization of the SHC concept, such undesired flow phenomena must be first predictable and second controllable. The lack of physical insight into the complex flow structure and suitable predictive tools might be two reasons why the SHC concept has not yet been realized despite its various advantages.

On that front, the problem of asymmetrically bounded swirling flows has rarely been explicitly investigated. And if so, similar to the reviewed literature in this section, understanding the role

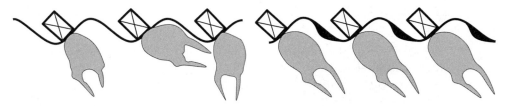

Figure 1.10: Schematic deflection and deformation of swirling flames in the SHC as a consequences of the asymmetric boundaries (left). Potential for flow control in the SHC by manipulation of the lateral boundaries (right).

of asymmetric boundaries was not explicitly the subject of investigations. Thus, the underlying physics of observed phenomena were not elucidated in those reports.

Within the framework of the present SHC study, a fundamental investigation of asymmetrically bounded swirling flows was performed by Schäfer (2018). It was based on a numerical study of two-dimensional rotating flows via the integral balance of angular momentum. A parametric study was carried out, which involved rectangular and circular wall geometries as well as symmetric and asymmetric confinements. It was found that the torque induced by the pressure field on a rectangular wall is generally higher than that on a circular wall. In the case of symmetrical confinement, the rotating flow does not exert a resulting torque on a confining circular wall. This is because the wall pressure forces are radially directed toward the rotational axis, and so they hold no torque lever arms. In contrast, in the case of a rectangular wall, the wall pressure forces, in general, have lever arms regarding the rotational axis. Hence, the rotating flow experiences a certain amount of reaction torque from a rectangular confining wall, which can significantly brake the rotating flow.

Furthermore, it is shown that in symmetric confinements the integral pressure torques are generally lower than those in the asymmetric confinements. This is because of the partial compensation of the pressure forces of the fronting walls in the symmetric cases. Moreover, it is shown that the torque induced by the wall pressure increases almost linearly with the rotational speed of the flow. An exception here is for the symmetrically confined circular walls, where the wall pressure torque vanishes.

Having such an insight for the 3D swirling flows, the flow can be manipulated by modification of the asymmetric boundaries. A motivating example would be the contouring of the combustor liner walls to guide the flow, as schematically depicted in Fig. 1.10. The advances in the additive manufacturing methods made possible the economical realization of such complex wall geometries in a gas turbine combustor.

The objectives and contributions of the present thesis are introduced in the next section.

1.3   Objectives and Contributions

The primary goal of this research study is to understand the characteristics of asymmetrically bounded swirling flames in an annular gas turbine combustor with angular air supply. This fundamental insight will be the basis of any attempt to treat and control the highly complex and nonuniform flow field in such an SHC.

The secondary goal is to investigate an approach to control the SHC flow based on the manipulation of the local pressure field in the primary combustion zone via contouring the liner walls. Such an approach might provide an important step toward attaining the technical maturity to employ this combustor in an aircraft engine.

For the fundamental studies of such a novel combustor concept, a numerical flow simulation is an unparalleled tool. It provides a vast amount of data for the aerothermal quantities of the 3D flow field. Thus, a global flow analysis based on integral methods is possible, which is impossible with the currently available experimental approaches.

Accordingly, to pursue the main goals of the present thesis, a framework of robust and effective numerical methods is established to explore the fundamental aerothermal features of the flow in the SHC. The following list summarizes the contributions of this work regarding the analytical–numerical tools:

1. Implementation of an existing combustion model in an open–source simulation code

2. Calibration and validation of the applied simulation approach based on experimental results

3. Development of a method for global analysis of the flow based on the integral balance of angular momentum

4. Derivation of a physically consistent technique to average the aerothermally nonuniform flow at the exit plane of the SHC

Numerous combustor models are developed and numerically analyzed. The focus was on gaining physical insight about the complex helical flow structure and testing the potential of the liner wall contouring to control the asymmetrically bounded swirling flows in the SHC. Thereby the following objectives are pursued

1. Derivation of scaling laws for the SHC concept based on similarity considerations

2. Parametric study of similar SHC models to find superior configurations regarding the aerothermal sensitivity of the flow

3. Elucidation of the causes of the flow nonuniformities in SHC with the aid of kinematic and dynamic flow analyses

4. Assessment of the liner wall contouring as a remedy to enhance the effectiveness of the SHC in utilizing the compressor angular momentum flow as well as a method to improve the aerothermal uniformity of the SHC exit flow

2 Theoretical Background

The objectives of this study are pursued by numerical methods. These are widely accepted tools in research and development of modern gas turbines. The methods for numerical simulation of fluid flows are denoted in general as Computational Fluid Dynamics (CFD). A numerical solution of the flow field can be obtained following a set of rules. These provide a discrete description of the governing equations and the solution domain, which are supplemented by appropriate boundary and initial flow conditions.

In this chapter, the flow equations and modeling approaches for describing turbulence and combustion are reviewed.

2.1 Computational Fluid Dynamics

The numerical methods applied in this study obey the laws of continuum mechanics. The continuum hypothesis is valid if the time- and length scales of the characteristic flow phenomena in the problems of interest are significantly larger than the microscopic scales of the matter. In this context, the macroscopic physics of the flow can be described as continuous functions of time and space (Bachelor, 1967). Accordingly, the combustion in a gas turbine can be described within the framework of continuum mechanics (Williams, 1985).

The combustor flow has been simulated with OpenFOAM®. It is an open–source C++ library for computational continuum mechanics developed by Weller et al. (1998).

In this section, the governing equations for gaseous reacting flows are briefly reviewed, and the modeling approaches to be applied for treating the turbulence and the premixed combustion are presented. For a detailed insight, the reader is referred to the textbooks by Anderson (1995) as well as Poinsot and Veynante (2012). Finally, the proposed numerical setup is validated by comparing the calculated with measured fields of velocity and temperature in a model gas turbine combustor.

2.1.1 Governing Equations

The framework of the continuum mechanics is established on the mass conservation, Newton's second law of motion as well as the first and second laws of thermodynamics (Aris, 2012). These are not provable empirical axioms. In the following, the integral formulations of the governing equations are introduced because they are plain and in accordance with human experience. The differential formulations, which are utilized in the majority of CFD problems, are formal derivations of the integral equations. The equations describing the motion of a fluid through an arbitrary control volume V, fixed in space and bounded by a closed surface A, can be written as (Hirsch, 2007):

- General transport equation of an intensive scalar ϕ

$$\underbrace{\frac{\partial}{\partial t} \int_V \rho\phi dV}_{time\ derivative} + \underbrace{\oint_A \rho\phi\, u \cdot dA}_{convective\ flux} + \underbrace{\oint_A \mathcal{J}_\phi \cdot dA}_{diffusive\ flux} = \underbrace{\int_V \Phi_V\, dV}_{volume\ source} + \underbrace{\oint_A \Phi_A \cdot dA}_{surface\ source} \qquad (2.1)$$

- Mass transport equation (Also called continuity equation)

$$\frac{\partial}{\partial t} \int_V \rho dV + \oint_A \rho u \cdot dA = 0 \qquad (2.2)$$

- Linear momentum transport equation (For viscous fluids known as Navier–Stokes equation)

$$\frac{\partial}{\partial t} \int_V \rho u dV + \oint_A \rho u (u \cdot dA) = \int_V \rho f_v\, dV + \oint_A \sigma \cdot dA \qquad (2.3)$$

- Angular momentum transport equation

$$\frac{\partial}{\partial t} \int_V r \times (\rho u) dV + \oint_A r \times \rho u (u \cdot dA) = \int_V r \times (\rho f_v) dV + \oint_A r \times (\sigma \cdot dA) \qquad (2.4)$$

- Energy transport equation

$$\frac{\partial}{\partial t} \int_V \rho h_t dV + \oint_A \rho h_t u \cdot dA = \int_V (\rho f_v \cdot u + Q_v) dV + \oint_A (\sigma \cdot u) \cdot dA - \oint_A q \cdot dA \qquad (2.5)$$

where

- dA is the surface element vector pointing alongside the outward normal,

- dV is the volume element,

- Φ_V is the volume source of the scalar ϕ,

- Φ_A is the surface source of the scalar ϕ,

- $\mathcal{J}_\phi$ is the diffusive flux of the scalar ϕ ,

- ρ is the density,

- u is the velocity vector,

- f_v is the vector of external volume forces,

- r is the position vector,

- σ is the stress tensor,

- h_t is the total specific enthalpy,

- Q_V represent all volume energy sources such as those by chemical reactions,

- q is the heat flux vector.

The transport equations are valid for any continuum. The resulting system of equations is, however, indeterminate. To close the system, additional information regarding the nature of the concerned continuum is required (e.g., perfect gas, incompressible fluid, dilatant material). For various types of continua, the corresponding data are summarized in the so-called constitutive relations. For Newtonian fluids in local thermodynamic equilibrium, the following relations can be applied (Bird et al., 2007):

- The equation of state

$$\rho = \rho(p, T) \tag{2.6}$$

- The enthalpy equation

$$h_t = h_t(u, p, T) \tag{2.7}$$

- Newton's law of viscosity

$$\boldsymbol{\sigma} = -p\boldsymbol{I} + \boldsymbol{\tau} = -p\boldsymbol{I} + \mu\left[\nabla\boldsymbol{u} + (\nabla\boldsymbol{u})^T - \frac{2}{3}(\nabla \cdot \boldsymbol{u})\boldsymbol{I}\right] \tag{2.8}$$

- Fick's law of diffusion

$$\mathcal{J}_\phi = -\mathcal{D}_m\rho\nabla\phi \tag{2.9}$$

- Fourier's law of heat conduction

$$q = -\lambda\nabla T \tag{2.10}$$

where

- $\boldsymbol{I}$ is the unit tensor,

- $\boldsymbol{\tau}$ is the viscous shear stress tensor,

- p is the pressure,

- T is the temperature,

- $\mathcal{D}_m$ is the mass diffusivity,

- λ is the heat conductivity.

For a given flow with available constitutive relations and appropriate specification of the boundary and initial conditions, the governing equations can be solved numerically. For numerical calculations, the space and time domain of the flow problem, as well as the governing equations, must be discretized. The former is obtained in this work by the Finite Volume Method (FVM), which is subdividing the space into a finite number of control volumes. This discretization includes the definition of the position of the points in the control volumes, where the solution is sought, as well as the description of the boundaries. The simulation time must be split into time steps, where the step length is limited by the Courant-Friedrichs-Lewy (CFL) number (Courant et al., 1928). The temporal, convective, diffusive, and source terms in the differential form of the governing equations are discretized to convert them into a system of algebraic equations. There are various discretization techniques for different terms of governing equations available. They directly affect the stability and accuracy of the solution. The applied discretization schemes in this work will be introduced in Chap.. 3. A comprehensive description of discretization schemes for FVM in OpenFOAM, provided by the corresponding implementations and error analyses, can be found in Jasak (1996).

2.1.2 Modeling of Turbulence

The flow regimes in engineering applications are predominantly turbulent. The velocity field in such flows is always unsteady, 3D, and fluctuates irregularly. The turbulent fluid motion exhibits a wide range of time- and length scales from large turbulent structures comparable with the dimension of the flow domain to smallest eddies characterized by the Kolmogorov microscales (Kolmogorov, 1941, 1962).

For the numerical solution of turbulent flow problems based on the FVM, there are three main approaches presently available. These are Direct Numerical Simulation (DNS), Large Eddy Simulation (LES), and calculations based on Reynolds Averaged Navier–Stokes equations (RANS). For the SHC investigations, the RANS technique is utilized. A brief introduction to the DNS, LES, and RANS approaches is given in Appendix A.1. A comprehensive review of simulation techniques for turbulent flows can be found in Ferziger and Peric (2012).

The standard k,ε-model is the most widely used approach to simulate the turbulent flow in the RANS context. In practical applications, it exhibits good results in the prediction of free-shear layer flows with relatively small pressure gradients. In addition, acceptable performance has been reported in the simulation of wall-bounded and internal flows with small mean pressure gradients (Bardina et al., 1997).

Nevertheless, the k,ε-model also has well-known deficiencies in the prediction of particular classes of flows. The inaccuracies mainly stem from the Boussinesq's eddy viscosity hypothesis and the turbulent dissipation rate ε. It is plausible that the turbulence becomes more isotropic at small scales. The large scales, at which the average quantities are defined, are not isotropic in many practical situations. A well-known case is the vortex breakdown phenomenon with turbulent structures in the central toroidal region, which are neither homogeneous nor isotropic. Such a nonreacting swirling flow in a model combustor was investigated by Jochmann (2007),

Jochmann et al. (2006) using an unsteady RANS method. Thereby, the prediction capability of the standard k,ε-model was compared with those of a Reynolds Stress Model (RSM) proposed by Speziale et al. (1991). They concluded that the roughly 30% increased computational effort of the calculations with the RSM was absolutely justified by the significant improvement of its simulation results compared with the standard k,ε-model.

Flow in the vicinity of an impermeable no-slip wall is another notable case, where the assumption of isotropic turbulence is violated. Strictly speaking, it is necessary to adapt both the eddy viscosity approximation and the resolution of computational mesh to resolve reasonably the near-wall details. The higher mesh resolution, however, can be compensated by the application of models based on "law of the wall" (Prandtl, 1932, von Kármán, 1930). The "wall functions" are principally simplified turbulence models, representing the near-wall characteristics of k and ε as introduced first by Launder and Spalding (1974). They assume a fully developed turbulent boundary layer in the vicinity of solid walls and couple the high gradients in this region with the rest of the domain.

In general, to provide a better description of the turbulent scales and the energy exchanges between them, the standard k,ε-model can be modified and extended for different flow conditions. The basic ideas among others are

- Calibrating the model by adapting the corresponding constants in the production and dissipation terms of Eq. (A.10) and Eq. (A.11).

- Modifying the production and dissipation terms in the transport equation for ε to account for the contribution of smaller scales to eddy viscosity (Shih et al., 1995, Yakhot et al., 1992).

- Deriving new algebraic formulations for eddy viscosity (Gatski and Speziale, 1993, Pope, 1975, Wallin and Johansson, 2000, Zhang et al., 1992).

- Splitting the turbulence spectrum into several zones with specific transport equations for the turbulent kinetic energy and the dissipation rate (Kim, 1991, Menter, 1994, Wilcox, 1988).

In Chap. 3 the nonreacting flow in a model combustor is simulated using LES and RANS approaches. For the latter, the RSM (Launder et al., 1975) and the standard k,ε-model (Launder and Sharma, 1974), as well as its most popular modifications, are applied. These are the RNG- by Yakhot et al. (1992), the Realizable- by Shih et al. (1995) and the SST-model by Menter (1994).

The benchmarking demonstrated that the Re-Normalization Group (RNG) k,ε-model with modeled near-wall region is a suitable choice for simulation of the SHC flow. This numerical packet exhibited the best compromise between the computational cost, the ease of setup and configuration, the stability of the solution procedure, as well as the accuracy of the predictions. Escue and Cui (2010) came to a similar conclusion in a numerical study of turbulent swirling flow inside a smooth straight pipe using the ANSYS FLUENT ® CFD program. For swirl numbers of up to unity, they found the RNG-k,ε-model performing similar to the RSM when comparing the predicted mean velocity profiles to the experimental data. In addition, Weller et al. (1998) demonstrated the comparable performance of the RNG-k,ε-model to other high-fidelity approaches in the introduction paper to OpenFOAM. They simulated the flow around a square prism using LES with different Sub Grid Scale (SGS) models, as well as RANS with RSM, standard k,ε, and RNG-k,ε-models.

The k,ε equations in the RNG variant are obtained by the RNG analysis of the Navier–Stokes equations Orszag et al. (1996). The constant values are different from those in the standard model. In addition, in the ε equation, there is an additional term. This term is an empirical calibrator, which is not derived from the RNG theory but mainly responsible for the better performance of RNG compared with the standard k,ε-model. A detailed description and benchmarking of the model for different flow cases can be found in Smith and Reynolds (1992), Smith and Woodruff (1998), Zhang and Orszag (1998).

2.1.3   Characteristic Turbulent Scales

A further feature of k,ε-models with particular advantages for combustion modeling and similarity analysis is the possibility of a consistent definition of turbulent length and time scales based on the eddy cascade hypothesis of Kolmogorov (1941). The relevant scales, which are used in Sec. 2.2 and Chap. 5, are defined in the following.

For homogeneous isotropic turbulence, it assumes a continuous transfer of kinetic energy from large eddies down to the smallest Kolmogorov eddies. At the smallest scales, the energy is dissipated by microscopic viscous forces. Phenomenologically, it can be explained as the continuous breakup of large vortex structures, like a waterfall, downward to smaller ones, until the smallest disappear due to viscous forces.

The energy spectrum of turbulent flow can be obtained by measuring the kinetic energy in the entire range of turbulent scales and plotting it against the inverse of a characteristic eddy size ℓ. In Fig. 2.1, a schematic representation of the turbulent kinetic energy spectrum is reproduced from Peters (2000), Pope (2000) and Hirsch (2007).

The interval between the integral scale $\ell_{\mathcal{I}}$ and the Kolmogorov scale ℓ_{κ} is called the inertial subrange. In this range, the energy is transferred locally from one eddy only to the eddy of the next smaller length scale. Because of this locality, the energy transfer rate, i.e., the kinetic energy per eddy turnover time, is independent of the length scale of the eddies, and thus constant within the inertial subrange. This scale independence of energy transfer is the reason for the prominent $-5/3$ slope of the kinetic energy function in the inertial subrange.

The "scale invariance" is the essential feature of the eddy cascade hypothesis, which holds strictly only for large Reynolds number flows. It is integrated into most of the standard turbulence models and satisfies the Reynolds number independence of the models in the large Reynolds number limit.

For large-scale eddies $\mathcal{L}$, which are typically characterized by the geometric boundary conditions of the flow, the kinetic energy scales with a power law between ℓ^{-2} and ℓ^{-4}. The spectrum

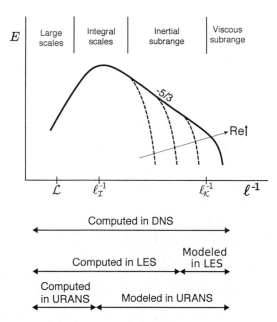

Figure 2.1: Log–log plot of the energy spectrum of turbulence as a function of its length scale, with an indication of the range of application of the DNS, LES, and RANS models.

exhibits a maximum at the integral scale $\ell_\mathcal{I}$, where the eddies contribute the most to the kinetic energy. The energy content of smaller eddies in the inertial subrange decreases with $\ell^{5/3}$. Beyond the Kolmogorov scale $\ell_\mathcal{K}$, in the so-called viscous subrange, the kinetic energy is rapidly dissipated by molecular mixing.

The integral length scale $\ell_\mathcal{I}$ can be defined by the normalized two-point velocity correlation function (Peters, 2000)

$$R(x,r) = \frac{\overline{u'(x,t)u'(x+r,t)}}{\sqrt{\overline{u'^2(x,t)}}\sqrt{\overline{u'^2(x+r,t)}}}$$

(2.11)

As depicted in Fig. 2.2, the $R(x,r)$ is a measure of the correlation of the velocity fluctuations u' at the location x with that in location $x+r$. It is equal to unity (maximal influence of neighbor location) in the limit of $r \to 0$ and decreases asymptotically to zero for $r \to \infty$. The integral length scale is then defined by

$$\ell_\mathcal{I} = \int_0^\infty R(x,r)dr$$

(2.12)

By definition, the value of $\ell_\mathcal{I}$ is located where the areas above and below the two-point velocity correlation function are equal. Accordingly, the integral length scale can be interpreted as the size from which the turbulent eddies are predominantly uncorrelated.

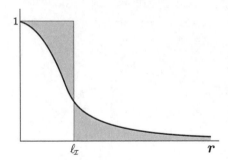

Figure 2.2: The normalized two-point velocity correlation for homogeneous isotropic turbulence. Reproduced from Herrmann (2001).

The velocity fluctuations associated with homogeneous isotropic turbulence are equal to each other in all spatial directions. Thus, the Reynolds averaged turbulent kinetic energy, as defined in Eq. (A.8), can be estimated as $k = 1.5\,\overline{u_{\mathcal{I}}^2}$, where the $u_{\mathcal{I}}$ represents the turnover velocity of integral eddies. Their turnover time $\ell_{\mathcal{I}}/u_{\mathcal{I}}$ is then given by definition from the eddy dissipation ε (the transfer of the kinetic energy per time). Accordingly, the integral time and length scales are given by

$$t_{\mathcal{I}} = \frac{k}{\varepsilon} \quad , \quad \ell_{\mathcal{I}} = \sqrt{2/3}\,\frac{k^{3/2}}{\varepsilon} \quad , \quad u_{\mathcal{I}} = \sqrt{2/3\,k} \qquad (2.13)$$

At the end of the energy cascade, the motion of small turbulent eddies is driven by viscous momentum diffusion as characterized by the kinematic viscosity ν. The Kolmogorov time and length scales, and so the turn over velocity of Kolmogorov eddies, can be derived by dimensional analysis

$$t_{\kappa} = \left(\frac{\nu}{\varepsilon}\right)^{1/2}, \quad \ell_{\kappa} = \left(\frac{\nu^3}{\varepsilon}\right)^{1/4}, \quad u_{\kappa} = (\nu\varepsilon)^{1/4} \qquad (2.14)$$

In combustion, molecular mixing plays an essential role. This takes place in thin layers typically smaller than the Kolmogorov scale. The time scales of the related chemical reactions are also smaller than almost all turbulent time scales. The geometric length scale of a combustor $\mathcal{L}$ can be estimated as one order of magnitude larger than the integral scale of the turbulent reacting flow in the flame tube. In engineering combustion, the Kolmogorov scales are typically over two orders of magnitude smaller than the integral scales (Peters, 2000). Hence, resolving all combustion processes occurring at the Kolmogorov scale requires more than three orders of magnitude between $\mathcal{L}$ and the smallest mesh size. In addition to that, a detailed description of the combustion process from the perspective of chemical kinetics requires solving of the transport equation for hundreds of chemical species reacting according to thousands of reaction mechanisms (Law, 2010, Smith et al., 2011).

The excessive computation demand for the description of reacting flows highlights the role of combustion modeling. In the following section, the modeling approach is introduced, which is used in this work for simulation of the turbulent premixed combustion.

2.2 Fundamentals of Turbulent Combustion

The central challenge in the prediction of turbulent combustion flows is the presence of a broad range of time and length scales. This range corresponds to various physical and chemical processes governing combustion, as well as the coupling between them, across different scales.

Bilger et al. (2005) have reviewed the various paradigms that have been established to describe the turbulent combustion phenomena. A pillar of almost all approaches is the "scale separation" between the turbulent and chemical scales to handle the complexity of the turbulent combustion.

The scale separation can be presumed in most applications of engineering combustion. It is implemented implicitly or explicitly in many practical turbulent combustion models (Fox, 2003). Accordingly, the high-temperature chemical reactions are often faster than all turbulent time scales. With the support of molecular diffusion, they take place in so-called reaction layers of width that is typically smaller than the Kolmogorov length scale ℓ_κ. With an exception for density changes, the reaction layers cannot exert feedback on the flow. On the other hand, in this context, the Kolmogorov eddies cannot penetrate into reaction layers and disrupt their structure. Thus, the scale invariance in the inertial subrange is not influenced by combustion. Hence, the time and length scales of combustion are separated from those of turbulence in the inertial subrange. It makes the mixing process in the inertial subrange independent of chemical processes and significantly simplifies modeling (Peters, 1997).

The RANS models, in their transient formulation, resolve the flow by sizes in the order of the integral length scale. Thus, within the context of scale separation, RANS impacts all scales in the inertial and viscous subrange arbitrarily. As a consequence, the inner structure of the flame cannot be resolved and only an average statistical position of the flame front can be predicted. The provided average flow field features scales that are much larger than the instantaneous flame thickness. In addition, the closure problem is particularly nonlinear for the reaction source terms in the species and some forms of the energy equations.

On the other hand, the large difference between the chemical and physical scales leads mathematically to stiffness of the solution matrices. Hence, by representing the flame with average scales, the performance of the computations can be significantly improved.

The limits mentioned above, however, set a framework for RANS combustion models that is quite restrictive and requires well-considered compromises. In the following, the different categories of turbulent combustion and the related modeling requirements are briefly reviewed. Subsequently, the applied modeling approach in this study and its theoretical background are introduced.

2.2.1 Turbulent Combustion Categories

Turbulent combustion can be divided into three main categories according to the condition of the fuel–oxidizer mixture prior to combustion: premixed, nonpremixed (diffusion), or partially premixed combustion. In premixed combustion, the fuel and oxidizer are homogeneously mixed on a molecular level before combustion takes place. Furthermore, the fuel-to-oxidizer ratio must lie between the flammability limits, which range typically from approximately $\varphi = 0.5$ to $\varphi = 1.5$. φ is the equivalence ratio defined as the fuel-to-oxidizer ratio in the unburned mixture normalized by that of a stoichiometric mixture. Once these conditions are satisfied, a heat source can ignite the mixture. Combustion then proceeds, and the flame front propagates through the mixture, which is the key feature of premixed combustion.

In nonpremixed combustion, fuel and oxidizer are separately injected into the combustion chamber. Consequently, nonpremixed flames do not propagate. They are located where fuel and oxidizer mix, and the rate of mixing controls the rate of reaction. Compared with premixed flames, the burners for nonpremixed combustion are more straightforward to design because a perfect mixing within the given geometrical limits is not required. Nonpremixed flames are also superior concerning operational safety because they do not exhibit flashback or autoignition in undesired locations.

The phenomenological description of nonpremixed and partially premixed combustion regimes, as utilized in many combustors, is more difficult than the premixed one. This primarily because the reactants must first be locally mixed before combustion can take place. On the other hand, the varying local mixing ratio of the reactants leads to fluctuating local heat release rates, which in turn give rise to more aerothermal inhomogeneity. Therefore, a nonpremixed flame is more sensitive to turbulence because the convective heat and mass transfer in combustion can be decisively influenced by turbulence. Thus, the description of the mixing process directly affects the accuracy of the reacting flow simulations. The second difficulty arises from the fact that nonpremixed flames do not exhibit well-defined characteristic scales. For example, a diffusion flame does not feature a propagation speed, and the flame thickness depends on the local flow conditions.

Following the primary assignment of the current study, the principal challenge lies in providing a reliable prediction and general understanding of the reacting swirling flow characteristics, when they are subjected to asymmetric boundaries in the SHC. In particular, at this preliminary stage of the SHC investigation, the prediction of species concentrations and the emission characteristics are beyond the scope. Factors that are much more important are an acceptable accuracy in reflecting the decisive flow features, the robustness of the solution algorithm, the ease of implementation of models and application of solvers, and all that at a reasonable computational cost. In that respect, the relevant data are the time-averaged velocity, pressure, and temperature fields. Accordingly, and without loss of generality regarding the aerothermal characteristics of swirl flames in the SHC, the combustor flows are simulated with the combustion of a homogeneous mixture of methane as the fuel and dry air as the oxidant.

Before selecting a modeling strategy for the simulation of turbulent premixed combustion, it is essential to have a comprehensive understanding of the mechanisms of turbulence–chemistry interaction in such combustion processes. The corresponding modeling approaches can be derived based on an analysis of the various time and length scales involved in specific combustion regimes.

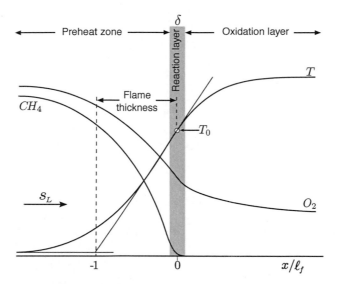

Figure 2.3: Schematic illustration of the structure of a stationary, laminar, premixed methane–air flame. Reproduced from Peters (1997).

2.2.2 Characteristic Scales in Premixed Combustion

The required chemical time and length scales for the classification of combustion regimes can be determined by studying the inner structure of flames. The structure of a laminar premixed flame is schematically depicted in Fig. 2.3. It was obtained by Peters and Williams (1987) from the analysis of a stationary, laminar, stoichiometric methane–air flame based on a four-step reduced mechanism. The flame architecture contains a chemically inert preheat zone with a thickness of the same order as the flame, followed by the reaction layer, where the fuel is consumed. This layer is responsible for sustaining the combustion process. For stoichiometric methane–air flames at ambient conditions, $p_a = 1e5\,Pa$, $T_a = 300K$, the temperature at the reaction layer is $T_0 = 1370K$ while the adiabatic flame temperature is $T_b = 2225K$ (Seshadri and Peters, 1990). The structure of the reaction layer can be disrupted if the small turbulent eddies penetrate into it. The flame would extinguish if the Kolmogorov eddies excessively transport heat and radicals out of the reaction layer. Therefore, the scale separation between turbulence and chemistry requires that the reaction layer ℓ_δ is thinner than the Kolmogorov length scale ℓ_κ. Subsequently, in the oxidation layer primarily carbon monoxide and hydrogen molecules are oxidized to carbon dioxide and water with minor influence on the flame characteristics Peters (1997).

The chemical length scale which is represented by the thickness of a laminar flame ℓ_f is defined as (Göttgens et al., 1992)

$$\ell_f = \left(\frac{T}{\partial T / \partial x} \right)_0 = \frac{(\lambda / c_p)_0}{(\rho\, s_L)_u} \tag{2.15}$$

Thereby, the subscripts 0 and u indicate that the values are evaluated at the reaction layer temperature T_0 and in the unburned gas, respectively. It is implicitly assumed that the mass diffusivities of all reactive species are the same and proportional to the thermal diffusivity. The representative flame diffusivity is then denoted by $\mathcal{D}_f$

$$\mathcal{D}_f = \frac{(\lambda/c_p)_0}{\rho_u} = s_L \ell_f \qquad (2.16)$$

The unstrained laminar burning velocity s_L is a thermochemical transport property defined as the velocity at which the flame front propagates into the unburned mixture. It depends primarily on the equivalence ratio φ, the temperature in the unburned mixture, and the pressure. s_L has been measured for various fuels over a wide range of these parameters (Gu et al., 2000, Law, 1993). It also can be calculated numerically using elementary or reduced reaction mechanisms and molecular transport properties (Mauß and Peters, 1993), or with empirical and semiempirical correlations as proposed by Gülder (1982, 1984, 1991).

The thickness of the reaction layer ℓ_δ is defined as

$$\ell_\delta = \delta \ell_f \qquad (2.17)$$

where Peters (1991) estimated the proportionality factor δ to vary from $\delta = 0.1$ at atmospheric pressure to $\delta = 0.03$ for pressures around $30e5\,Pa$.

The characteristic chemical time scale of a laminar flame is then given with

$$t_f = \frac{\ell_f}{s_L} = \frac{\mathcal{D}_f}{s_L^2} \qquad (2.18)$$

2.2.3   Regime Diagram for Premixed Turbulent Combustion

The phenomenological analysis of the coupled physical-chemical combustion processes is often summarized in the so-called "regime diagram." Thereby, various regimes are identified and delineated by dimensionless numbers regarding ratios of length and velocity scales. The regime diagrams are mainly based on intuitive arguments and introduce orders of magnitudes rather than precise demonstrations. The derivation is also based on strong assumptions of turbulence. Nevertheless, characterizing the flow via such diagrams is the first essential step toward building or selecting an appropriate turbulent combustion model. Such diagrams have been proposed by Abdel-Gayed et al. (1989), Borghi (1985), Bray (1980), Peters (1999), Poinsot et al. (1991) and Williams (1985) among others.

The nondimensional numbers, which are often utilized to construct such regime diagrams are defined below.

Turbulent Reynolds Number

The Reynolds number, Re, is an overall measure for the inertial forces relative to the diffusion forces. It is based on global scales and mean velocities. Similarly, a local turbulent Reynolds number can be defined with the turbulent scales. It compares the turbulent momentum diffusion with the viscous momentum diffusion. Assuming a Schmidt number $Sc = \nu/\mathcal{D}_m$ of unity and using Eq. (2.16), the turbulent Reynolds number is defined as

$$Re_t = \frac{\nu_t}{\nu} = \frac{u_\mathcal{I} \ell_\mathcal{I}}{\nu} = \frac{u_\mathcal{I} \ell_\mathcal{I}}{s_L \ell_f} \tag{2.19}$$

The Re_t provides a more precise description of the local flow property because the integral length scale $\ell_\mathcal{I}$, and the related eddy turning velocity $u_\mathcal{I}$, are likely to vary at different locations of the considered flow domain. It is also a measure of the relation of the turbulent integral and Kolmogorov length scales as discussed in Sec. 2.1.3

$$\frac{\ell_\mathcal{I}}{\ell_\kappa} \sim Re_t^{3/4} \tag{2.20}$$

In technical problems, the turbulent Reynolds numbers Re_t are typically one until two orders of magnitudes lower than the flow Reynolds number $Re = U\mathcal{L}/\nu$, considering $u_\mathcal{I}/U \approx 0.01 \cdots 0.2$ and $\ell_\mathcal{I}/\mathcal{L} \approx 1 \cdots 0.5$.

Turbulent Damköhler Number

The Damköhler number is defined as the ratio of the chemical reaction rate to a characteristic convective or diffusive mass transport rate (Damköhler, 1940). Analogously, the turbulent Damköhler number is defined as the ratio of the integral turbulent time scale to the chemical time scale t_f

$$Da_t = \frac{t_\mathcal{I}}{t_f} = \frac{s_L \ell_\mathcal{I}}{u_\mathcal{I} \ell_f} \tag{2.21}$$

A Damköhler number smaller than unity corresponds to fast turbulence compared with the chemical reaction. Reciprocally, $Da_t > 1$, as most common in gas turbine combustion, indicates fast chemistry, and thus a determining role of the turbulence.

Karlovitz Number

The Karlovitz number relates the characteristic chemical scales to the turbulent Kolmogorov scales

$$Ka = \frac{t_f}{t_\kappa} = \left(\frac{\ell_f}{\ell_\kappa}\right)^2 = \left(\frac{u_\kappa}{s_L}\right)^2 \tag{2.22}$$

It indicates whether the smallest eddies can modify the structure of the flame front.

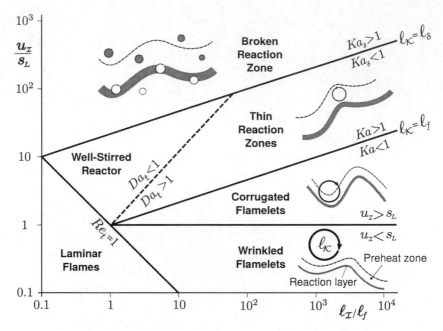

Figure 2.4: Regime diagram for premixed turbulent combustion.

Referring to the role of the reaction layer for the stability of a premixed flame, a second Karlovitz number can be defined as the ratio of the reaction layer thickness to the Kolmogorov length scale. Using Eq. (2.15), the Ka_δ is given as

$$Ka_\delta = \left(\frac{\ell_\delta}{\ell_\kappa}\right)^2 = \delta^2 Ka \qquad (2.23)$$

It indicates whether the Kolmogorov eddies are small enough to enter the reactive layer.

A regime diagram for turbulent premixed combustion is presented in Fig. 2.4. It is derived from Peters (1999) and supplemented with schematic illustrations of flame–vortex interactions. The regions are identified in terms of length ($\ell_{\mathcal{I}}/\ell_f$) and velocity ($u_{\mathcal{I}}/s_L$) ratios and delimited by the lines of constant Reynolds, Karlovitz, and Damköhler numbers. The relation between the axis variables may be expressed as

$$\frac{u_{\mathcal{I}}}{s_L} = Ka^{2/3}\left(\frac{\ell_{\mathcal{I}}}{\ell_f}\right)^{1/3} = Re_t\left(\frac{\ell_{\mathcal{I}}}{\ell_f}\right)^{-1} = Da_t^{-1}\left(\frac{\ell_{\mathcal{I}}}{\ell_f}\right) \qquad (2.24)$$

Before discussing the different regimes, the underlying assumptions for the construction of this diagram are reviewed below. The relations are based on the scaling laws applicable to homogeneous isotropic turbulence at large Reynolds numbers. The boundaries are adiabatic, and the mass diffusivity is the same for all reactive scalars and equal to the thermal and viscous momentum diffusivity (Lewis $Le = \mathcal{D}_T/\mathcal{D}_m$ and Schmidt numbers $Sc = \nu/\mathcal{D}_m$ equal to unity).

Nevertheless, the diagram provides an order of magnitude estimation of the flame–turbulence interaction. The various regimes in Fig. 2.4 including laminar flames, wrinkled and corrugated flamelets, thin reaction zones, well-stirred reactor, and broken reaction zones are discussed next.

Laminar Flames

When $Re_t < 1$, which is the condition in the lower-left corner of the diagram, the basic flow is laminar, and the combustion lies on laminar flame structures. As stated before, the underlying analyses constructing the regime diagram are strictly valid in the limit of large Reynolds numbers, which corresponds to a region sufficiently removed from the line $Re_t = 1$ toward the upper right of the diagram.

Flamelet Regime

The flamelet term corresponds to a continuous thin flame front having the internal structure of a one-dimensional laminar flame front. While the turbulent time scales are larger than the chemical time scales ($Da_t > 1$), the Kolmogorov eddies are still larger than the laminar flame thickness ($Ka < 1$). Thus, the turbulent eddies cannot perturb the internal structure of the laminar flame. As a consequence, the flame front remains thin, and the interaction between flame and turbulence is purely kinematic. The flamelet regime may be subdivided into two regions.

The first one is the "wrinkled flamelet" regime, which corresponds to rather weak turbulent structures compared with the flame propagation. It is bounded by the condition $u_t/s_L < 1$. The deformations of the flame front due to turbulence are quickly compensated by the advancement of the front, so that only small wrinkling of the flame front may be observed. This regime is irrelevant for the gas turbine combustion because in the related applications the intensity of the turbulent flow is typically very high.

The second flamelet regime, the so-called "corrugated flamelet," is bounded by $u_t/s_L > 1$ and $Ka < 1$. The fluctuation velocities of the integral eddies are larger than the laminar burning velocity, which corrugate the flame front. $Ka < 1$ implies that the smallest eddies are still larger than the laminar flame thickness, cf Eq. (2.15), so that they cannot significantly modify the inner structure of the flame. The topology begins to change so that occasionally pockets of fresh and burned gas can be observed. However, the interaction between turbulence and the laminar flame structure is still extensively kinematic, so that the chemical processes within the flame structure are essentially unmodified.

Thin Reaction Zones

This regime is bounded by $Ka > 1$ and $Ka_\delta < 1$. In contrast to the corrugated flamelet regime, the Kolmogorov scales are smaller than the flame thickness and able to penetrate into the preheat zone. On the other hand, they are still larger than the thickness of the reaction layer ($\ell_\delta < \ell_\kappa < \ell_f$), so that they cannot disrupt the chemical kinetics of the flame. Hence, the presence of the

Kolmogorov eddies in the preheat zone enhances the transport of chemical species and heat between the reaction layer and the unburned gas. Nevertheless, turbulence does not essentially influence the chemical reactions. As a result, the diffusion within the preheat zone, and so in the flame thickness, is increased. The flame front is still continuous and can be described as an ensemble of "thickened flamelets." For this regime, turbulence and combustion cannot be completely dissociated, and the concept of the scale separation reaches its limit. It should be noted that because of the enhanced mixing, higher volumetric heat release, and shorter combustion times, technical combustion devices are predominantly operated in the thin reaction zones regime.

Well-Stirred Reactor

This regime is defined for scenarios when the flow exhibits a fairly high turbulence intensity so that the turbulent motions have shorter characteristic times than the chemical reaction time $Da_t < 1$, but the small scale eddies are unable to disrupt the reaction layer of the flame front $Ka_\delta < 1$. In this regime, mixing is fast, and the overall reaction rate is limited by chemistry.

Broken Reaction Zones

This is the regime where the Kolmogorov eddies are smaller than the thickness of the reaction layer ($Ka_\delta > 1$) and able to enter it. Turbulence perturbs the reaction layer and the internal structure of the flame. The chemical reactions can be locally suppressed owing to the loss of heat from the reaction layer, followed by temperature decrease and degeneration of reactive radicals. This is illustrated in Fig. 2.4 by gray Kolmogorov vortices transporting heat and radicals out of the flame, and white vortices transporting cold and nonreacting fluid into the reaction layer. Consequently, the flame speed may be reduced, or the flame extinguishes locally. This process is classically described as quenching. The combustion stops locally and the fresh reactants and products interdiffuse without burning. In this situation, the description of the reacting flow is not possible with standard flamelet approaches. The quenching phenomenon determines the flamelet limit. It separates two fundamentally different structures of the turbulent premixed flames, and the corresponding description and modeling approaches (Poinsot et al., 1991).

2.3   Combustion Modeling

As mentioned in Sec. 2.2, a major challenge in the modeling of turbulent combustion is the partic-ipation of numerous species in chemical reactions through thousands of elementary mechanisms. Thus, the simplification of reaction mechanisms is a mandatory task in combustion modeling.

In premixed combustion, the flow consists of two regions: The unburned reactants and the products from combustion. The extent of the combustion can be described by a progress variable taking values between 0 (unburned) and 1 (fully burned), and the transition between these values marks the flame front. Hence, the entire reaction mechanism can be represented by a single reactive scalar, the reaction progress variable c. It is linked directly to the physical properties of the gas such as a normalized temperature or product mass fraction

$$c = \frac{T - T_u}{T_b - T_u} \quad \text{or} \quad c = \frac{Y_p}{Y_{p,b}} \tag{2.25}$$

This implies a one-step irreversible reaction $E \rightarrow P$, from reactants to products, and an associated rise of temperature due to heat release. The Favre-weighted transport equation for $\tilde{c}$ is given by

$$\frac{\partial \bar{\rho} \tilde{c}}{\partial t} + \nabla \cdot (\bar{\rho} \tilde{v} \tilde{c}) = \nabla \cdot \left(\bar{\rho} \mathcal{D}_f \nabla \tilde{c} - \bar{\rho} \widetilde{u'' c''} \right) + \bar{\omega}_c \tag{2.26}$$

with the flame diffusivity $\mathcal{D}_f$ and the mean reaction rate $\bar{\omega}_c$. Similar to the Reynolds stresses, the turbulent scalar flux $\bar{\rho} \widetilde{u'' c''}$ is caused by the averaging procedure. Analogous to the Boussinesq Eddy Viscosity assumption Eq. (A.8), the "gradient diffusion" hypothesis can be applied to model this term

$$\widetilde{u_x'' c''} = -\mathcal{D}_t \frac{\partial \tilde{c}}{\partial x} \tag{2.27}$$

The turbulent mass diffusivity $\mathcal{D}_t$ is related to the turbulent momentum diffusivity v_t with the turbulent Schmidt number Sc_t

$$Sc_t = \frac{v_t}{\mathcal{D}_t} \tag{2.28}$$

where Sc_t can be assumed to be equal to unity.

If the flow in the vicinity of the flame is dominated by thermal expansion instead of turbulence, typically for $u_I/s_L < 3$, the turbulent transport may become "countergradient"

$$\widetilde{u_x'' c''} / (\partial \tilde{c} / \partial x) > 0 \tag{2.29}$$

In general, this case is not relevant for gas turbine combustion with large turbulent intensities. A more detailed discussion on this topic can be found in Peters (2000) and Poinsot and Veynante (2012).

The transport equation for the reaction progress variable can also be reformulated to

$$\frac{\partial \bar{\rho} \tilde{c}}{\partial t} + \nabla \cdot (\bar{\rho} \tilde{v} \tilde{c} - \bar{\rho} \mathcal{D}_e \nabla \tilde{c}) = \bar{\omega}_c \tag{2.30}$$

with effective diffusivity $\mathcal{D}_e$ calculated from

$$\mathcal{D}_e = \mathcal{D}_f + \mathcal{D}_t \tag{2.31}$$

2.3.1 Mean Reaction Rate

The closure of Eq. (2.30) still requires the modeling of the chemical source term $\bar{\omega}_c$, which is highly nonlinear due to the chemical reaction rate equations and the Arrhenius law (Arrhenius, 1889). This term must reflect the turbulence–flame interactions like wrinkling, curvature, and thickening, as discussed in Sec. 2.2. The quenching of the flame at very high turbulence conditions may also be reproduced by approaching the $\bar{\omega}_c$ to zero in that situation.

The prominent models in the literature for describing the mean reaction rate $\bar{\omega}_c$ are among others the Eddy Breakup (Spalding, 1971), Eddy Dissipation(Magnussen and Hjertager, 1981), the Bray-Moss-Libby (Bray and Moss, 1977), the Flame Surface Density (Trouvé and Poinsot, 1994), and the Turbulent Flame Speed Closure (TFC). The latter is employed in this thesis and introduced in more detail below.

2.3.2 Turbulent Flame Speed Closure

Zimont and Lipatnikov (1995) originally proposed the TFC model. It estimates the mean reaction rate based on the evolution of the flame front $|\nabla \tilde{c}|$, propagating with the turbulent flame speed s_T

$$\bar{\omega}_c = \rho_u s_T |\nabla \tilde{c}| \tag{2.32}$$

Zimont and Lipatnikov (1995) theoretically deduced the following expression for s_T

$$s_T = \frac{1}{2} u_\mathcal{I} Da_t^{1/4} \tag{2.33}$$

The turbulent flame speed s_T, however, is not a well-defined quantity. Experimental investigations demonstrate a broad scatter of values depending on various parameters such as chemistry characteristics, turbulence scales, and initial flow conditions (Gouldin, 1996). Based on the experimental data of Abdel-Gayed et al. (1985, 1989), Bradley (1992), Gülder (1991) and the theoretical analysis of Yakhot et al. (1992), a general model for the turbulent flame speed s_T may be given by

$$\frac{s_T}{s_L} = 1 + c_{st} \left(\frac{u_\mathcal{I}}{s_L} \right)^n \tag{2.34}$$

where c_{st} and n are two model constants dependent on the combustion regime (Poinsot and Veynante, 2012).

2.3.3 Schmid TFC Model

In the present study, a modified TFC model is used, which was theoretically derived by Schmid (1995). This model showed good correlations with measurements of s_T in almost all regions of the combustion regime diagram as shown by Schmid et al. (1998), Zhang (2014), Zhang et al.

(2009, 2013) and Trapp (2014). An exception is the quenching effects in the case of a very high turbulence intensity in the broken reaction zone because the chemistry is taken into account in the Schmid model only by the unstrained laminar flame speed.

In the Schmid model, an RMS of chemical and turbulent time scales is employed as the characteristic time scale for the description of the mean reaction rate. Using the Kolmogorov-Petrowski-Piscounov theorem (Kolmogorov et al., 1937), the following relation was proposed for the mean reaction rate $\bar{\omega}_c$

$$\bar{\omega}_c = \frac{c_{kpp}\rho_u}{\sqrt{t_\mathcal{I}^2 + t_f^2}} = c_{kpp}\rho_u \frac{s_T^2}{u_\mathcal{I}\ell_\mathcal{I}} \tilde{c}(1 - \tilde{c}) \tag{2.35}$$

where c_{kpp} is the model constant, and the turbulent flame speed is given as

$$\frac{s_T}{s_L} = 1 + \frac{u_\mathcal{I}}{s_L} \left(1 + Da_t^{-2}\right)^{-\frac{1}{4}} \tag{2.36}$$

where the turbulent Damköhler number and the related scales are obtained with Eqs. (2.13, 2.15, 2.21). In this work, the laminar flame speed is calculated with the empirical correlations of Gülder (1984).

In the TFC models of Zimont and Schmid, the mean rate $\bar{\omega}_c$ is estimated with different physical aspects. The former with tracking the flame surface and the latter based on a global time scale for the heat release.

The turbulent flame speed s_T as given by Eq. (2.36), exhibits a linear rise with $u_\mathcal{I}$ in the flamelet regime ($Da_t \gg 1$). Consequently, the mean reaction rate tends to be inversely proportional to the integral time scale

$$Da_t \gg 1 \quad \rightarrow \quad s_T \approx u_\mathcal{I} \quad \rightarrow \quad \bar{\omega}_c \sim 1/t_\mathcal{I} \tag{2.37}$$

If the turbulent integral time scales are in the same order or smaller than the chemical time scales ($Da_t < 1$), the s_T rises degressively with the turbulence intensity. Thus, the chemistry is the rate-determining factor

$$Da_t < 1 \quad \rightarrow \quad s_T \approx u_\mathcal{I} Da_t^{1/2} \quad \rightarrow \quad \bar{\omega}_c \sim 1/t_f \tag{2.38}$$

The mass fractions of reactants and products in the flow domain can be calculated with the reaction progress variable, which is available as the solution of Eq. (2.30). The heat release as the source term of the energy equation is then calculated according to the concentration of the species in the mixture. The associated enthalpy of the mixture is determined with the temperature-dependent heat capacity c_p using the JANAF tables (Chase et al., 1975). Finally, the temperature is calculated in an iterative algorithm with the energy conservation equation Eq. (2.5).

2.3.4   Pressure Dependence of the Mean Reaction Rate

The combustion in gas turbines must take place at elevated pressures to achieve a higher thermal load. The pressure dependence of the mean reaction rate, as defined in Eq. (2.35), results from the variation of the density of the unburned gases, as well as from the pressure dependence of the turbulent and the chemical time scales.

The Reynolds number $Re = \rho U \mathcal{L}/\mu$, as the measure for the turbulent character of the flow, increases with the pressure. This is mainly due to the higher density of the fluid at elevated pressures. The velocity, as evident in the Navier–Stokes equation, changes mainly with the gradient of the pressure. The minor pressure dependence of the dynamic viscosity of real gases can be neglected as well (Kestin and Leidenfrost, 1959).

On the other hand, the flow in the atmospheric combustion systems, where the mean reaction rate models are typically validated, exhibits Reynolds numbers of typically $\mathcal{O}(4)$ and above. As experimentally shown by Mydlarski and Warhaft (1998), this range satisfies the large Reynolds number limit for the independence of the integral turbulent scales (see Sec. 2.1.3 and Fig. 2.1). Thus, the integral turbulent time scale can be estimated as invariant with respect to the elevation of the pressure.

Hence, the pressure dependence of the mean reaction rate can be estimated, to the first order, as controlled by the chemical time scale. This can be analytically derived, based on the theories of the chemical reaction kinetics as

$$t_f \sim p^{-n+1} \tag{2.39}$$

where n is the overall order of the chemical reaction (Law, 2010, Turns, 2012). The n varies for oxidation of different hydrocarbons between $1 < n < 2$. At stoichiometric conditions, for methane is $n \approx 1$ and for n-octane is $n \approx 1.75$ (Westbrook and Dryer, 1981). Accordingly, for stoichiometric fuel–air mixtures the turbulent Damköhler number correlates with the pressure as

$$Da_t = \frac{t_{\mathcal{I}}}{t_f} \sim \frac{p^0}{p^{-n+1}} \sim p^{n-1} \tag{2.40}$$

However, in addition to the fuel type, also the equivalence ratio of the fuel–air mixture influences the pressure dependence of the chemical reaction kinetics significantly. Particularly, in the case of lean combustion the excessive air molecules are not reflected in the overall reaction order. This exhibit the practical limitation of the analytical relations given in Eqs. (2.39,2.40).

With Eq. (2.18), the pressure dependence of the chemical time scale can be estimated via that of the flame diffusivity and the laminar flame speed.

Based on Eq. (2.15), the pressure influences the flame diffusivity mainly via the density because the pressure effect on the heat conductivity λ and the specific heat capacity c_p can be neglected, so that

$$\mathcal{D}_f \sim \rho_u^{-1} \sim p^{-1} \tag{2.41}$$

For the laminar flame speed, the experimental findings by Andrews and Bradley (1972), Gu et al. (2000), Kobayashi et al. (1996, 1997, 1998), Metghalchi and Keck (1980), among others, suggest for a wide range of fuels, equivalence ratios, and pressures of up to 50 bars

$$s_L \sim p^{-\gamma} \quad \text{with} \quad \gamma \approx 0.1 \cdot \cdot 0.5 \tag{2.42}$$

Hence, for different hydrocarbons and fuel–air mixing ratios, an empirical relation for the pressure dependence of the chemical time scale can be given by

$$t_f \sim \frac{\mathcal{D}_f}{s_L^2} \sim \frac{p^{-1}}{p^{-2\gamma}} \sim p^{2\gamma-1} \tag{2.43}$$

For lean methane–air mixtures, as employed in the present work, a value of $\gamma = 0.5$ is in accordance both with the experimental findings of Kobayashi et al. (1997) and the analytic solutions of Mauß and Peters (1993). The latter is based on the thermal flame theory of Zel'dovich and Frank-Kamenetzki (1938). Accordingly, the chemical time scale t_f and the turbulent Damköhler number Da_t, exceptionally for the lean methane–air flames can be considered as pressure independent.

It must be here again noted, that the estimation of the pressure independence of the chemical time scales is only justified for lean methane–air mixtures. Particularly, for combustion of methane in rich conditions or combustion of long-chain hydrocarbons, the value of γ becomes smaller, and thus, a pressure dependence of the chemical time scale and turbulent Damköhler number arises.

With $Da_t \sim p^0$ in Eq. (2.36), and in the limit of $u_{\mathcal{I}} \gg s_L$ for technically relevant regimes, the turbulent flame speed can also be estimated as pressure independent $s_T \sim u_{\mathcal{I}} \sim p^0$. Applying this in Eq. (2.35) yields

$$\bar{\omega}_c \sim \rho_u \frac{u_{\mathcal{I}}^2}{u_{\mathcal{I}} \ell_{\mathcal{I}}} \tilde{c} (1 - \tilde{c}) \sim \rho_u \sim p \tag{2.44}$$

which is in accordance with the experimental observations (Lipatnikov and Chomiak, 2002). As a conclusion, the turbulent mean reaction rate of the Schmid model exhibits, for the combustion of lean methane–air mixtures as applied in the present work, a plausible trend in pressure dependence.

3 Validation of the Numerical Method

The turbulence and combustion models presented in Chap. 2 are validated in this chapter. For this purpose, the simulation results of the reacting and nonreacting flows in an experimentally well-investigated model gas turbine combustor are compared with the LDV and Raman spectroscopy measurements. Several numerical setups in the RANS and LES frameworks are benchmarked with a focus on the quality of simulations regarding the mean and transient flow quantities and the associated computational costs.

As a result, for the prediction of the SHC flow, a superior CFD framework is configured. It features robust boundary and consistent initial conditions. Moreover, it delivers a sufficient level of flow description at a reasonable numerical effort, which made it possible to perform a comprehensive parametric study of the SHC concept.

3.1 Experimental Setup

The so-called "Preccinsta combustor" is a derivation of an industrial design. The flow in this model combustor has been experimentally investigated in the German Aerospace Center (Meier et al., 2007, Weigand et al., 2005, 2007). The experimental rig, the swirler, and the computational domain are depicted schematically in Fig. 3.1.

Dry air at atmospheric pressure and a temperature of approximately $320K$ is fed via a plenum through a swirler with 12 radial passages to the burner nozzle. The total air mass flow rate is $12g/s$ and the fuel gas (methane) is injected into the airflow through small holes within the swirler. The high momentum of fuel jets combined with the high velocity in the swirler channels ensure an excellent mixing before entering the combustion chamber, so that perfectly premixed combustion can be assumed. A central cone-shaped hub supports the flame stabilization through the recirculation zone and controls the flame position. Accordingly, in the experiments, a blue V-shaped turbulent flame was observed anchored at the central hub as shown in Fig. 3.1.

The combustion chamber consists of large quartz windows creating a square section of $85mm \times 85mm$ and a height of $h = 114mm$. The exit of the combustion chamber is conically shaped leading to a central cylindrical exhaust pipe with a contraction ratio of approx. 0.2. The large windows enabled good optical access to almost the whole flame zone. The mean and RMS velocities were measured using LDV for the nonreacting and reacting flows. The measurement accuracy of $\pm 2\%$ is reported by Meier et al. (2007).

3.2 Numerical Setup

The following phenomena are neglected in the simulations undertaken in this study

- Compressibility and acoustic interactions

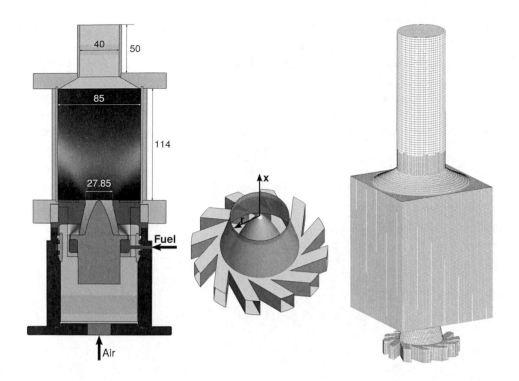

Figure 3.1: Schematic of the experimental burner setup with a photograph of a stable flame (left) reproduced from Meier et al. (2007). The radial swirler with a central hub for flame stabilization (center). The abstracted computational domain, which contains the swirler, the combustion chamber, and an extended exhaust pipe (right).

- Body forces

- Thermal radiation

Neglecting compressibility effects is justified for flows with low Mach numbers. Thereby, it is presumed that thermodynamic variables such as density, temperature, enthalpy, and entropy are decoupled from pressure variations. For solving the governing equations, a fully implicit pressure-correction method is applied. Here, the so-called PISO formulation (Pressure-Implicit with Splitting of Operators) of Issa (1986) is employed.

Second-order schemes are employed for discretization of the space and time domains within the FVM framework of the present study. Nevertheless, to treat the irregularities of the mesh and preserve the boundedness of the solution, the accuracy requirements have to be relaxed in certain parts of the solution. Therefore, each term of the governing equations is treated individually by an appropriate discretization technique. The general form of the temporal, convective and diffusive terms are briefly introduced below.

For the temporal discretization, a mixed implicit method is applied, which is a linear combination of 70% second-order "Crank–Nicolson" and 30% first-order accurate "Euler" method. The convective flux terms are discretized with "Blended Differencing" schemes, consisting of an even portion of second-order "Central Differencing" and first-order "Upwind Differencing" to preserve both boundedness and accuracy of the solution. The diffusion terms were discretized with a "Central Differencing" scheme, where the second-order accuracy for nonorthogonal parts of the mesh is maintained by adding an explicit nonorthogonality correction as described by Jasak (1996).

The flow is fully turbulent with a Reynolds number of approximately $Re \approx 31,000$ calculated with the nozzle diameter of $D = 27.85mm$, and with the cold flow viscosity and velocity at the nozzle exit.

For validation studies, a stable operating point with an fuel–air equivalence ratio of $\varphi = 0.75$ is simulated. At this operating point, the flame does not exhibit thermoacoustic instabilities so that the low-Mach-number approximation is applicable. The Karlovitz number is estimated to be 35, which points out that the combustion takes place in the thin reaction zones of the regime diagram as introduced in Fig. 2.4.

The physical boundary conditions introduced in Sec. 3.1 are numerically implemented as:

- **Inlet:** The velocity boundaries at the inlets are fixed values (Dirichlet type) giving the specified mass flow. The pressure inlet boundaries are set as zero gradient (Neumann type) to preserve consistency with the velocity boundaries (Hirsch, 2007). With Eq. (2.13), the turbulent variables are set to construct a turbulence intensity of 5% of the bulk inlet velocity and an integral length scale half as large as the characteristic inlet dimension. In the swirler channels, turbulence further develops spontaneously and attains the correct magnitude at the inlet of the combustion chamber. All other scalars are prescribed as fixed values, which represent their average physical state at the inlet.

- **Outlet:** The boundary conditions at the outlet should be specified in such a way that the overall mass balance for the computational domain is satisfied. Because of the velocity–pressure coupling, the mass balance can be guaranteed by coordinating the outlet with the inlet boundaries. For the given inlet boundary as above, the velocity is set to be a zero gradient, and the pressure as a fixed value type. On the other hand, in the case of reactive flow simulations, the fixed value condition for pressure may give rise to unphysical reflections of pressure waves at the outlet boundary. This purely numerical phenomenon can lead to spurious flow oscillations inside the computational domain and consequently destabilization of the simulation. It can be prevented by assigning a so-called nonreflecting pressure boundary condition as proposed by Poinsot and Lelef (1992). A specific formulation of this kind of boundary condition, denoted as *waveTransmissive* in OpenFOAM, is used for the simulations. Turbulence and all other scalars are set as zero-gradient type.

- **Wall:** The walls are modeled as adiabatic and impermeable with the no-slip condition. Thus, the velocity of the fluid on the wall is equal to zero. The turbulent flow variables in the near-wall region are calculated using standard wall functions. The boundary conditions for pressure, temperature, and other scalars are set as zero-gradient type.

The computational grids are generated with the OpenFOAM native tool *snappyHexMesh*. The interior of the flow domains contains purely hexahedra elements. By approaching the boundaries, the hexa-elements are iteratively refined and snapped, as unstructured split-hex elements, to the boundary surfaces (OpenCFD, 2015). The walls are overall provided with a single layer of prismatic elements, with their heights set to achieve a dimensionless wall-distance of $30 < y^+ < 200$. Unlike the computational domains used in Galpin et al. (2008), Moureau et al. (2007), Roux et al. (2005), the plenum and the atmosphere outside the exhaust pipe are not included in the present study as shown in Fig. 3.1. In those works, the acoustic interaction between the turbulent flame and the system boundaries could be critical. Thus, a fully compressible Navier–Stokes solver was used. For such solvers, a larger computational domain is advantageous, with inlet and outlet far away from the combustion chamber. However, with the low-Mach-number approach, as applied in this study, compressibility effects are avoided. Nevertheless, the exhaust pipe was extended to three times longer as in the experimental configuration. In addition, the mesh was coarsened at the extended outlet by a factor of two. Thus, a buffer zone is created, which ensures a uniform and undisturbed flow at the outlet boundary.

The RANS/LES computations were carried out in the High-Performance Computing center at KIT (HPC, 2019). The time steps of transient simulations are chosen so that the maximum CFL number in RANS calculations remained below 0.8 and in LES below 0.4. The computed physical time in all simulations of the Preccinsta combustor corresponds to approximately nine flow throughputs in the computational domain. From that, three throughput times are simulated for establishing the flow and the subsequent six for gathering the statistical values of the flow field.

3.3   Comparison of Predicted and Measured Flow Fields

An extensive measurement database is available, which makes the Preccinsta combustor suitable for benchmarking and validation of turbulence and combustion models. In the LES context, several models were employed to predict the flow, including the artificially thickened flame (ATF) model (Roux et al., 2005), the flame surface density (FSD) model (Wang et al., 2016), an extended level-set algorithm (Moureau et al., 2007), a presumed probability density function (PDF) (Wang et al., 2014), a presumed beta-shape PDF (Galpin et al., 2008), a Filtered Tabulated Chemistry (Fiorina et al., 2010), as well as a combined presumed PDF and FSD method (Lecocq et al., 2011).

In the following, the performances of different turbulence and combustion models in RANS, as well as LES frameworks, are compared with the results of Roux et al. (2005) and Wang et al. (2016).

3.3.1   Mesh Independence Study

The first step for the assessment of different turbulence models is to provide a computational mesh. Foremost, the discretization must ensure a sufficient spatial resolution in the mainstream and near-wall regions. In the RANS framework, a proper mesh can be determined by a mesh independence study. Accordingly, starting from a fairly coarse mesh, the spatial resolution of the grids is increased until the results do not improve or even start to worsen. The latter would occur due to the excess of turbulent eddy viscosity in a too-fine mesh.

In contrast, in the LES framework the modeled subgrid scale (SGS) turbulence is directly related to the grid size. Thus, increasing the mesh resolution leads to a decrease of the modeled turbulence, and if the grid is fine enough to resolve the Kolmogorov scales the LES transforms into the DNS. Pope (2000) suggests that a computational grid for a reasonable LES should be fine enough to resolve at least 80% of the turbulent scales in the energy spectrum as illustrated in Fig. 2.1. This criterion is the basis for the selection of a mesh for the LES calculations later in this section.

For the RANS grid dependence tests, three different grid resolutions were investigated. The total cell numbers were approximately 140k (coarse), 1M (medium), and 8M (fine). Associated cutouts of these meshes from their inlet regions are shown in Fig. 3.2. Higher resolutions are obtained from the previous coarser mesh by an equal splitting of the grids in each space direction. The mean edge length of a cell in the medium mesh is 1mm. The x-axis points downstream with its origin at the peak of the nozzle cone. The computations for the grid independence study are carried out with the RNG-k,ε-model.

Fig. 3.3 shows the vector field of the time-averaged velocity with isolines of the streamwise velocity component. A large IRZ, evidenced by the isoline of zero streamwise velocity, develops on the chamber axis. The IRZ begins at the peak of the cone-shaped hub and is extended up to $x \approx 70mm$. Other sizable Outer Recirculation Zones (ORZs) exist in the dead-water region due to the sudden-expansion at the inlet of the chamber.

The time-averaged velocity profiles, Fig. 3.4, and the corresponding RMS fluctuations, Fig. 3.5, are predicted with different mesh resolutions and compared with the LDV measurements at various sections of the combustion chamber.

The comparison of time-averaged profiles shows good agreement of all simulations and experimental results in the vicinity of the swirl nozzle. Further downstream from $x = 15mm$ the simulation results start to deviate from the measurements both in the peak velocity and the lateral spread of the IRZ, which is more obvious for the coarse mesh. The medium and fine meshes still exhibit reasonable results for streamwise and azimuthal velocity components in all sections. The fine mesh performs slightly better in predicting the peak velocities and in near-wall regions ($r/D \approx -1.5$ and 1.5). It should be noted that the meshes in this study are not provided with boundary layer refinements to resolve the turbulent near-wall structures. Compared with the available LDV measurements, however, the near-wall velocities for the medium and fine meshes are fairly well predicted with wall functions.

The RMS velocities in Fig. 3.5 are computed from the unsteady RANS solution. For $x = 1.5mm$,

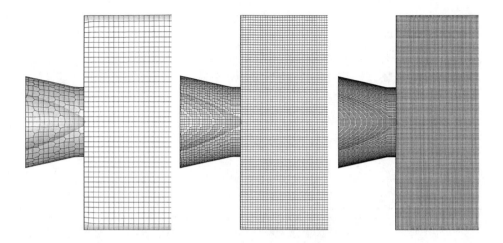

Figure 3.2: Different computational grids with a decreasing mean element size of 2mm for coarse (left), 1mm for medium (center), and 0.5mm for fine mesh (right).

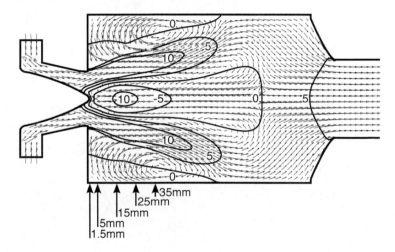

Figure 3.3: Nonreacting flow. Vector plot of the time-averaged velocity field predicted with RANS-RNG-k,ε on the medium mesh. Solid isolines correspond to the labeled time-averaged streamwise velocities in [m/s]. Filled arrows point to the x-position of the LDV measurements.

the LDV measurements of the RMS velocities are only available for $0 < r/D < 1.5$ and are mirrored for the other half of the chamber in all upcoming figures. For the medium mesh, the modeled parts of the turbulence are added on top of the resolved velocity fluctuations and indicated by bars. The modeled turbulence is very small at the inlet of the chamber ($x = 1.5mm$) compared with the resolved velocity fluctuations. It increases gradually as more turbulence is developed due to shear flows. Further downstream, the modeled part takes a more significant portion, on average about 70% of the total turbulent fluctuations.

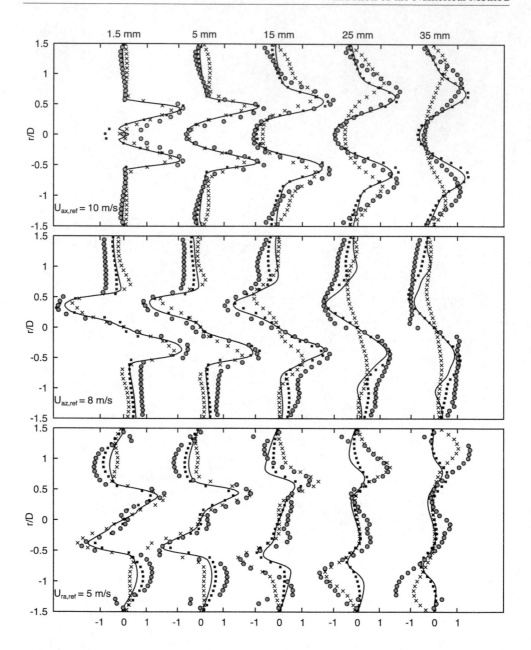

Figure 3.4: Nonreacting flow. Time-averaged velocity profiles. Axial (top), azimuthal (middle), and radial (bottom) velocity components are predicted via RANS-RNG-k,ε with the coarse (×), the medium (—), as well as the fine mesh (■) and measured with LDV (O). The velocity components are normalized with $U_{ax} = 10\,m/s$, $U_{az} = 8\,m/s$, $U_{ra} = 5\,m/s$.

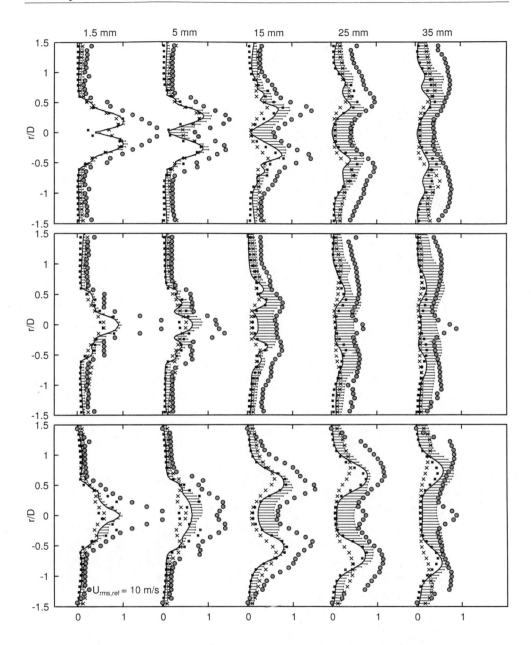

Figure 3.5: Nonreacting flow. Resolved RMS velocity profiles. Axial (top), azimuthal (middle), and radial (bottom) RMS velocity components are predicted via RANS-RNG-k,ε with the coarse ($\times$), the medium (—), as well as the fine mesh (■), and measured with LDV (O). The RMS velocity components are normalized with $U_{rms} = 10\,m/s$. The bars indicate the modeled turbulent fluctuations calculated using $\sqrt{2/3 \cdot k}$ on the medium mesh.

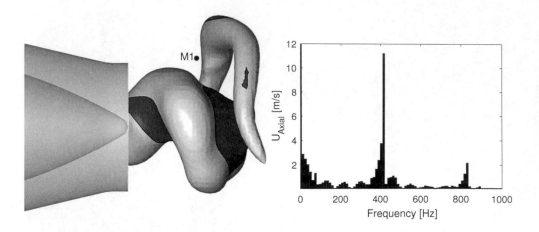

Figure 3.6: Nonreacting flow. Visualization of the 420 Hz Precessing Vortex Core (PVC) using an isosurface of low pressure (left). Dark and bright contours refer to negative and positive streamwise velocities, respectively. Spectra of streamwise velocity fluctuations (right) at the monitor point $M1(x = r = 0.54D)$.

Around the swirler axis, the levels of measured, but also predicted, RMS velocities are significantly high. These intensive fluctuations are due to a well-known large scale hydrodynamic structure in swirling flows, called the PVC (Syred et al., 1975, 1997). Measurements of pressure fluctuations with microphones revealed a dominant frequency around $510Hz$. From the RANS calculations, a dominant frequency of about $420Hz$ is obtained by fast Fourier transformation of the streamwise velocity field at a monitor point in the vicinity of the swirler.

The PVC is visualized in Fig. 3.6 with a low pressure isosurface and illustrated with the spectrum of the streamwise velocity obtained from resolved transient RANS calculations. All three meshes predicted approximately the same frequency and amplitude for the large scale fluctuations due to the PVC. Apparently, the large scale unsteady structures like the transient hydrodynamic instabilities, as well as the high-velocity gradients present due to the entrainment of the swirling jets in the chamber, can be well captured even on the coarse mesh. However, further downstream where the mixing through small scale turbulence becomes the determining factor, the coarse mesh suffers from the lack of modeled turbulent momentum diffusion.

The benefit of the fine mesh in capturing the peak velocities is marginal and does not justify the additional computational effort. It required approximately 16 times more resources than the medium mesh. This is due to the eight times bigger mesh and doubled number of time steps to maintain the same CFL condition. Hence, the medium mesh (1M cells) is chosen for further validations because it exhibits the best compromise between computational effort and accuracy of the results.

A comprehensive benchmark of other prominent RANS and LES turbulence models in predicting the cold flow in the Preccinsta burner can be found in Appendix A.2.

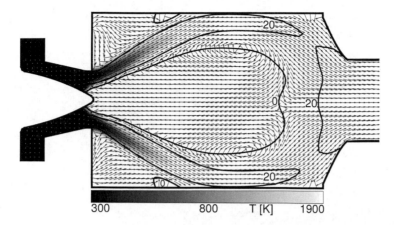

300   800   T [K]   1900

Figure 3.7: Reacting flow. Vector plot of the time-averaged velocity field predicted with RANS-RNG-k,ε-Schmid on the medium mesh. Solid isolines correspond to the labeled time-averaged streamwise velocities in [m/s]. The contour plot refers to the time-averaged temperature field.

3.3.2   Combustion Model

In this section, two reacting flow fields for fuel–air equivalence ratios of $\varphi = 0.75$ and 0.83 are predicted with RANS and LES. The RANS predictions are carried out with the RNG-k,ε turbulence model and the Schmid combustion model as described in Sec. 2.3. For the LES cases, the SV-SGS turbulence model is used (Nicoud et al., 2011). The interaction of the flame front and turbulence is modeled by two well-known models, the Partially Stirred Reactor (PaSR) model and the ATF model. The former model is available in OpenFOAM (Chen, 1997, Kärrholm et al., 2008), whereas the latter is implemented into the code by Schneider (2014) based on the works of Colin et al. (2000) and Angelberger et al. (2000). For combustion of the methane–air mixture, a reduced two-step mechanism (TSM) is considered, which takes six species into account ($CH_4, CO, CO_2, O_2, H_2O, N_2$). The reaction rates are then given by Arrhenius expressions as described by Selle et al. (2004).

Fig. 3.7 shows the vector field of the time-averaged velocity and isolines of the streamwise velocity plotted over contours of the time-averaged temperature field. Compared with the nonreacting flow field presented in Fig. 3.3, the IRZ is enlarged mainly in the radial but also in the axial extent. The expansion of the reacting gases due to the heat release leads to an overall higher level of streamwise velocities. This flow acceleration can be well recognized on the isolines of $U_{ax} = 20m/s$ for the swirling jet and the flow at the inlet of the exhaust pipe. Similar situations in the nonreacting flow are characterized by isolines of lower streamwise velocities $U_{ax} = 5m/s$. Moreover, the lateral expansion of the swirling jet suppresses the ORZ. Consequently, in the reacting case, only a small recirculation zone at the chamber walls is observed. This seems to be due to the flow separation right before the swirling jet impinges the chamber walls.

Similar to the nonreacting flow analysis, for the equivalence ratio $\varphi = 0.75$ the quantitative assessment of the simulation tools is carried out by comparison of the predicted velocity fields with the LDV measurements. For the reacting flow case $\varphi = 0.83$, the temperature fields are compared with those obtained by laser Raman scattering by Meier et al. (2007). For a cross comparison with other combustion models and CFD codes, the LES data extracted from Roux et al. (2005) and Wang et al. (2016) are shown as well. In the former work, the flame–turbulence interaction is treated with an ATF method, whereas the latter employed an FSD method in combination with a dynamic-Smagorinsky (DS) SGS model, as proposed by Moin et al. (1991). Both of the cross-compared simulations used, similar to the reacting LES cases of this study, a reduced TSM scheme for the chemistry modeling.

Fig. 3.8 shows the radial profiles of the time-averaged velocity vectors at different axial distances downstream from the Preccinsta burner. At first glance, all simulation tools can reproduce the overall features of the mean flow field.The LES tools are superior to RANS in predicting the peak as well as the near-wall flow situations. For later investigations of the SHC, however, of primary importance is the ability to predict the characteristic lateral expansion of the reacting swirling jet. The RANS simulations demonstrated this capability by the well-captured profiles of the radial velocities.

The evolution of the velocity profiles along the streamwise direction with the increasing radial distance between the velocity peaks is an indication of the radial expansion of the swirling jet. Thereby, the balance equation for angular momentum explains the lower azimuthal velocities of the radially more expanded combustive compared with the cold swirling jet. Accordingly, the larger is the radius of the swirling jet the lower is the azimuthal velocity to maintain a constant flux of angular momentum.

A consequence of combustion is the damping of the PVC. As demonstrated in Fig. 3.9, the radial and azimuthal RMS velocities are almost halved in the vicinity of the swirler. Furthermore, in Fig. 3.8, at the center of the profiles, the velocities and their gradients are both close to zero. This also supports the absence of the PVC. Consequently, the large coherent turbulent structure disappears, which was visualized for the nonreacting flow in Fig. 3.6. The combustion-induced higher dilatation and viscosity of the burned gases seem to impair the mechanisms that lead to the formation of this hydrodynamic instability flow phenomenon. The suppression of PVC by combustion was already observed in other swirl-combustors (Selle et al., 2004), however, it is not a general mechanism as discussed by Syred (2006).

For the turbulent fluctuations, the LES results exhibit very good agreement with the experiments. Particularly at $x = 1.5mm$ and $x = 5mm$, two shear layers can be distinguished. An outer one between the swirling jet and the surrounding fluid, and another inner shear layer between the jet and the IRZ. The cross-comparison in Fig. 3.9 shows that the present LES tools are superior in resolving the RMS fluctuations at the inner and outer shear layers of the swirling jet, as well as at the near-wall regions.

Compared with LES, RANS is significantly weaker in predicting the turbulent activities of the reacting flow. The reason can be elucidated by comparison of Fig. 3.9 with the nonreacting RANS results, Fig. 3.5. In contrast to the nonreacting flow, the resolved RMS velocities by

RANS are almost negligible until $x = 15mm$. The reason is the absence of the PVC and the associated large scale and low-frequency turbulence, which could be well captured by RANS. Instead, the velocity fluctuations associated with the inner and outer shear layers dominate the turbulent activities in the reacting flow. It seems that RANS cannot resolve or model these local high-frequency fluctuations, whereas the LES results reflect them very well.

Further downstream, the fresh reactants are progressively consumed, and simultaneously the shear layers become weaker. The latter can be recognized in Fig. 3.7 by diverging isolines of low and high streamwise velocities. This divergence indicates smaller velocity gradients and thus milder turbulent shearing with larger eddies in downstream stations, which again can be better captured by RANS. Hence, both the resolved and modeled turbulence by RANS increases while the flow evolves downstream. However, the magnitudes of the predicted RMS velocities deviate far from those of LES and LDV.

Despite the deviations in predicting the absolute values of velocity fluctuations, the RANS could precisely reflect the positions of their peaks. This is an important feature, which is necessary for the correct prediction of the flame position. For the combustion models, which maintain the heat release as a function of turbulent mixing, the effect of turbulence with a proper model constant can be augmented. Hence, for a certain operating point, the heat-release can be calibrated. The RANS of reacting flows in this study with the Schmid model were calibrated by $c_{kkp} = 1.5$ for simulations of the Preccinsta combustor.

For the high-pressure simulations of the SHC ($T_{in} = 820K$, $p = 31$bar), there are no comparable measurement data available. In this case, notwithstanding the reasonable pressure dependency of the Schmid model, the mean reaction rate must be further augmented by $c_{kpp} = 2.5$ to attain a complete burnout. The mean aerodynamic features of the flow at the elevated pressure in the SHC, as discussed in Sec. 2.3.4, are expected to be as well predicted as those in the atmospheric Preccinsta model combustor.

Radial profiles of the time-averaged temperature and its RMS values for $\varphi = 0.83$ are shown in Fig. 3.10. The overall agreement of both RANS and LES results with measured mean temperatures is reasonably good. Beyond $x = 40mm$ the fresh gases are nearly fully consumed as can be deduced from the mean temperature contours in Fig. 3.7. Hence, combustion is almost finished, and the radial profile of the mean temperature becomes flat. The spread-angle of the measured flame is broader than the predicted one in all simulations. It can be recognized through the radial shift of peaks downstream $x = 15mm$. Furthermore, both RANS and LES overestimate the temperatures in the outer region. The reason for these similar characteristics of both numerical tools is the imposed adiabatic boundary condition on the wall, which ignores the heat losses to the ambient. Thus, the near-wall discrepancies in temperature profiles are not a consequence of the combustion models. However, for combustor simulations at similar engine operating points, adiabatic boundaries are more reasonable because the realistic heat losses are marginal.

Close to the burner inlet, the magnitudes of temperature fluctuations reach up to 25% of the adiabatic flame temperature. These tremendous oscillations are due to the strong intermittency and flapping of the flame in this zone. Even though the magnitudes of the peaks are overall slightly underpredicted, the LES results exhibit good agreements with the measured RMS temperatures.

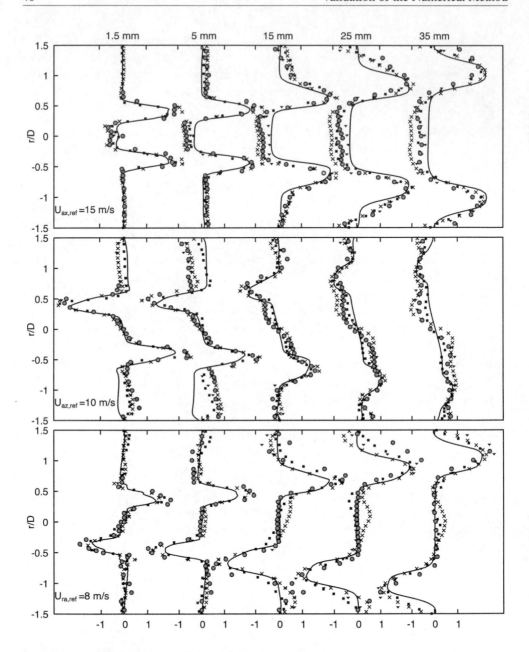

Figure 3.8: Reacting flow. Time-averaged velocity profiles. Axial (top), azimuthal (middle), and radial (bottom) velocity components are predicted via RANS-RNG-k,ε-Schmid (—), LES-SV-ATF-TSM ($\times$), LES-SV-PaSR-TSM ($\blacktriangledown$), LES-AVBP-WALE-ATF-TSM ($\blacksquare$) compared with LDV measurements ($\bigcirc$). The velocity components are normalized with $U_{ax} = 15\,m/s$, $U_{az} = 10\,m/s$, $U_{ra} = 8\,m/s$.

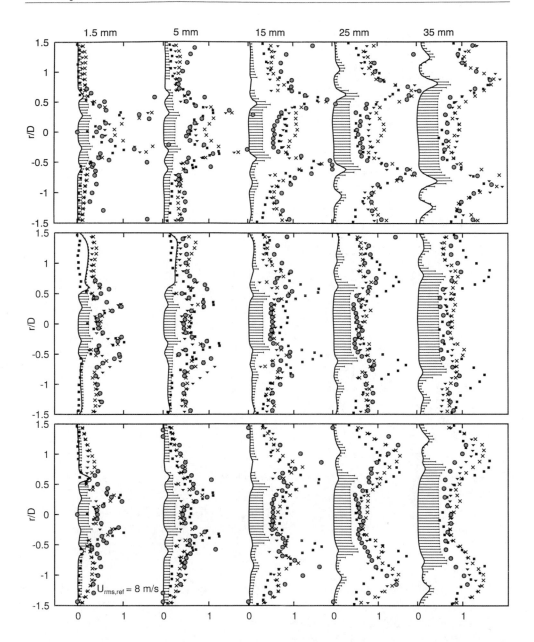

Figure 3.9: Reacting flow. Resolved RMS velocity profiles. Axial (top), azimuthal (middle), and radial (bottom) RMS velocity components are predicted via RANS-RNG-k,ε-Schmid (—) with bars indicating the modeled part calculated with $\sqrt{2/3 \cdot k}$, LES-SV-ATF-TSM ($\times$), LES SV-PaSR-TSM ($\blacktriangledown$), LES-AVBP-WALE-ATF-TSM ($\blacksquare$) compared with LDV measurements ($\bigcirc$). The RSM velocity components are normalized with $U_{rms} = 8\,m/s$.

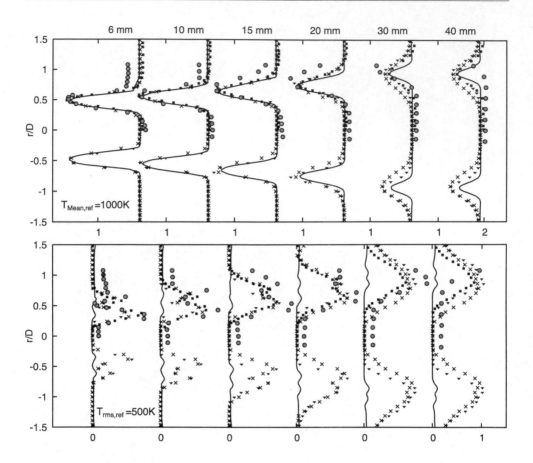

Figure 3.10: Reacting flow with an equivalence ratio of $\varphi = 0.83$. Time-averaged (top) and RMS temperature profiles (bottom). The flow fields are predicted via RANS-RNG-k,ε-Schmid (—), LES-SV-ATF-TSM ($\times$), LES-SV-PaSR-TSM ($\blacktriangledown$), LES-DS-FSD-TSM ($\blacksquare$) compared with measurements via laser Raman scattering ($\bigcirc$). The temperatures are normalized with $T_{mean} = 1000\,K$, $T_{rms} = 500\,K$.

On the other hand, the RANS utterly failed to predict the temperature fluctuations. However, via calibration of the combustion model constants, the effect of fluctuations in the *mean* velocity and temperature fields could be modeled. Hence, the accuracy of the predicted mean temperatures with RANS is comparable to those with LES. This model *manipulation*, however, evidently does not affect the transient thermal characteristics of the reacting flow predicted by the RANS method.

The instantaneous and time-averaged flame structures obtained by RANS-Schmid, LES-ATF, and LES-PaSR models are compared in Fig. 3.11. Thereby, isosurfaces of temperature $T = 1100\,K$ are plotted in the reacting flow with $\varphi = 0.75$. The instantaneous flame fronts obtained by LES are highly wrinkled with a significantly larger surface, whereas the RANS surface exhibits a

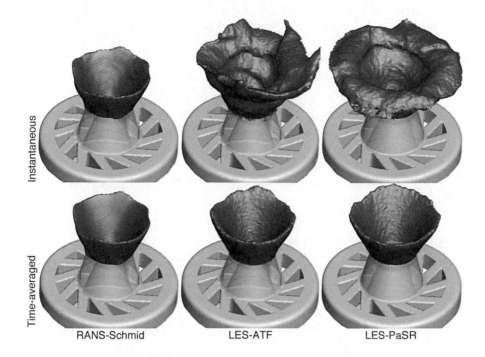

Figure 3.11: Iso-surface of instantaneous and time-averaged temperature at 1100K predicted by RANS-RNG-k,ε-Schmid (left), LES-SV-ATF-TSM (center), and LES-SV-PaSR-TSM (right).

smooth conical shape very similar to its time-averaged counterpart. On the other hand, the shape and size of all time-averaged flame fronts, obtained by RANS and LES, are very similar. This confirms the quantitative results of Fig. 3.10.

In selecting an appropriate numerical approach, it must be considered that capturing the finer turbulent structures and higher flow gradients in LES is at a computational cost of around 50 times higher than RANS. The application of LES is inevitable for flow studies, where the transient nature of the flames is the critical physical phenomenon. Prominent applications are the investigations of emission or thermoacoustic characteristics of gas turbine combustors.

As discussed in the introductory chapter, the objectives of the presented study need to be addressed by integral assessments of the combustor flow field. For that, the time-averaged flow quantities are the primary data of interest. For this purpose, the RANS method offers the right and sufficient level of description of the turbulent reacting flow. The decisive advantage of RANS is, however, that its numerical cost is about two percent of an equivalent LES. Hence, the numerical simulations regarding the parametric studies of the SHC are carried out within the RANS framework.

3.4   Asymmetrically Bounded Swirling Flow

Up to this point, it is demonstrated that RANS is a suitable method for prediction of the principal features of nonreacting and reacting swirling flows when considering the time-averaged data and symmetric boundary conditions. However, as discussed in Sec. 1.2.2, the response of a swirling flow to asymmetric boundaries can be highly nonlinear in space. If besides, the corresponding response was highly nonlinear in time, then the time-averaged flow quantities would be essentially dependent on such transient effects. In that case, the application of the RANS method must be carefully examined.

In particular, in the case of the SHC, the asymmetric confinement of the recirculation bubbles might substantially change the structure of the swirling flows. To examine the effect of transient nonlinear flow phenomena, the nonreacting flow is predicted in an asymmetrically confined Preccinsta-like combustor by both RANS and LES. The corresponding instantaneous and time-averaged results are exhibited in Fig. 3.12.

It can be seen that LES, by its nature, resolves much more small scale turbulent structures on the fine mesh with 8M elements. On the other hand, RANS on the medium mesh with 1M elements can only capture the large scale transient flow features such as the PVC, which can be identified in the wave structures on the swirling jet.

The comparison of time-averaged flow fields reveals that, apparently, the nonlinear temporal effects do not play a determining role in forming the overall structure of the flow. The differences between the RANS and LES prediction of the mean flow in an asymmetrically bounded Preccinsta swirler are within limits, which also have been seen when it was symmetrically bounded. Primarily, the location of the recirculation bubble, the opening angle of the swirling jet and the overall velocity levels are similarly predicted by RANS and LES.

In the case of combustion and asymmetric boundaries, additional aerothermal effects can become relevant. As an example, the contact of swirling flames with the cooled sidewalls might cause flame anchoring or quenching on the walls. However, the liner cooling and heat transfer effects, as well as the quenching phenomenon, are omitted in the present study. Thus, it can be assumed that RANS, with the introduced TFC combustion model, can predict the mean reacting flow quantities of an asymmetrically bounded swirling flame as well as those of a symmetrically bounded one.

The methods for an integral analysis of the flow are introduced in the following sections.

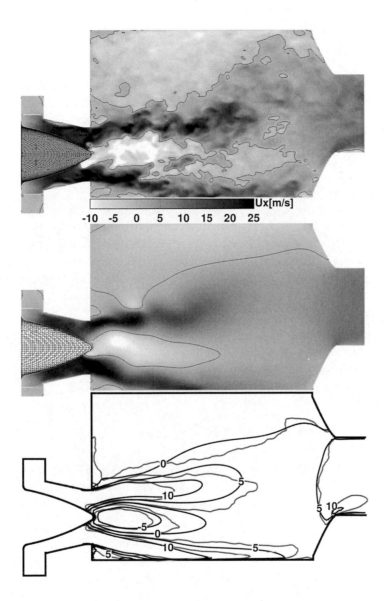

Figure 3.12: Nonreacting flow in the Preccinsta combustor with an asymmetrically confined swirler. Contour plots of instantaneous streamwise velocity with isolines of the zero value predicted with LES-SV on the fine mesh (top) and RANS-RNG-k,ε on the medium mesh (center). Isolines of time-averaged streamwise velocities (bottom). Black lines predicted with RANS and grey lines with LES.

4 Global Flow Analysis

Up to this point, a robust and efficient numerical framework for the prediction of reacting flow fields of the SHC is configured. The asymmetrically bounded swirling flows establish a complex helical flow structure in the SHC. Thus, the insight gained through a phenomenological flow analysis will be marginal and not enough for developing concepts to control the SHC flow. For that, it is primarily necessary to identify and quantify the physical mechanism driving the flow.

Furthermore, a reasonable quantity is required for the assessment of different SHC configurations in a parametric study. In the case of the SHC, the mean flow angle at the outlet of the flame tube and in front of the NGV appears to be a convenient parameter. It can be directly compared with the tilting angle of the burner axes at the inlet of the flame tube, and so can be deemed as a measure for the effectiveness of a certain SHC configuration in the utilization of angular momentum of the compressor flow. The flow at the exit plane of the SHC, however, can be totally nonuniform with respect to the velocity and temperature fields. In such a flow condition, the use of conventional area-weighted or mass-flow-weighted averaging procedures can be associated with a significant violation of the energy conservation law and the balance equations for angular and linear momentum flows.

In this chapter, as the next step, methods are developed for a global and consistent analysis of the predicted combustor flow. These are, on the one hand, a tool for the assessment of the flow field based on the integral balance of angular momentum, and on the other hand, a physically consistent procedure for the calculation of the spatial averages of flow quantities such as the flow angle in a gas turbine combustor.

4.1 Integral Balance of Angular Momentum

The main feature of the SHC concept, as introduced in Sec. 1.2.1, is the utilization of angular momentum flow arising from the compressor. Accordingly, measuring and tracking the amount of this flow quantity is essential for the flow analysis within the SHC. This can be done through the evaluation of the integral balance of angular momentum.

The objective is to elucidate the mechanisms determining the course of angular momentum flow through the flame tube from the inlet to the outlet. Thereby, the torques can be identified, which alter the flux of angular momentum. Moreover, the critical zones along the combustor where the forces become maximum can be localized. This insight would facilitate a targeted controlling of the flow, and thus, the improvement of the combustor performance.

To that end, the balance equation for angular momentum, Eq. (2.4), is rewritten for the case of Newtonian fluids (Eq. 2.8) as

$$\frac{\partial}{\partial t} \int_V \boldsymbol{r} \times (\rho \boldsymbol{u}) dV + \oint_A \boldsymbol{r} \times \rho \boldsymbol{u} (\boldsymbol{u} \cdot d\boldsymbol{A}) = \oint_A \boldsymbol{r} \times ((-p\boldsymbol{I} + \boldsymbol{\tau}) \cdot d\boldsymbol{A}) + \int_V \boldsymbol{r} \times (\rho \boldsymbol{g}) dV \quad (4.1)$$

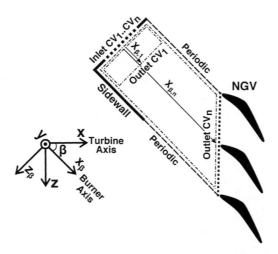

Figure 4.1: Successive CVs CV$_1$..CV$_n$ for the integral balance of angular momentum of the SHC flow. Dashed-dotted lines are representative CV, dashed lines periodic boundaries, and dotted lines are inlets and the combustor outlet.

The integral terms in Eq. (4.1) can be calculated from simulation results. After three initializing cycles and an averaging time for a further nine flow cycles, the computed time-averaged quantities, as well as their RMS values, reached stationary values. On that account, in the absence of gravitation forces Eq. (4.1) can be simplified and evaluated on a control volume (CV) in the SHC as

$$\underbrace{\int_{A_{out,cylics}} r \times \rho u (u \cdot dA)}_{L_{out}} + \underbrace{\int_{A_{in}} r \times \rho u (u \cdot dA)}_{L_{in}} = \underbrace{\int_{A_{in,out,periodics,walls}} r \times (-p I \cdot dA)}_{L_p} + \underbrace{\int_{A_{in,out,periodics,walls}} r \times (\tau \cdot dA)}_{L_\tau} \quad (4.2)$$

A single flame tube segment is schematically depicted with associated CVs and the coordinate system in Fig. 4.1. In that context, L_{in} refers to the initial flux of angular momentum, which results from the flow through the burners into a CV. L_{out} is the sum of angular momentum flows through the periodic and the outlet surfaces of the CV. L_p is the sum of the pressure torques and L_τ refers to the torques induced by all viscous forces.

The integral values for L_{out}, L_p, and L_τ in Eq. (4.2) are evaluated on successively larger CVs CV$_1$ until CV$_n$. The integral values are then plotted over the position x_β along the burner axis between the inlet and some stations along the streamwise direction in the flame tube. Thereby, the upstream boundary of CVs remains fixed at the inlet, while the downstream boundary moves along the burner axis until the combustor outlet. The side boundaries are then given by the partial sector surfaces, i.e., the periodic sides and the inner/outer liners. The course of integral values indicates zones of influence while the value of the final CV, here the CV$_n$, gives the overall contributions to the balance of the combustor.

With the proposed integral flow analysis, the evolution of different terms (advection, pressure, diffusion) of the balance equation for angular momentum can be monitored, and so reveal their contribution in forming the flow structure. This tool will be applied to evaluate quantitatively the complex asymmetric SHC flow. The gained fundamental insight is then used to find and invent solutions to improve the uniformity of the flow and increase the efficiency of the SHC in terms of higher exit mean flow angles.

In the following section, an averaging method is proposed, which is suitable for calculating mean quantities from the aerothermally highly inhomogeneous flow at the combustor outlet.

4.2   Physically Consistent Averaging Method

For the representation of the mean flow quantities in fluid mechanics, area- or mass-weighted averages are frequently employed. Despite their ease of use, these methods are arbitrary and induce physical inconsistencies into the system. In a comprehensive work, Pianko and Wazelt (1983) concluded that "no uniform flow exists which simultaneously matches all the significant stream fluxes, aerothermodynamics and geometric parameters of a nonuniform flow." Consequently, the averaging method of the flow quantities such as total pressure or temperature is neither arbitrary nor optional. Instead, the averaging must be adapted for the specific purpose, for which the representative mean quantity is sought. In the following, a method is derived that is suitable for physically consistent averaging of the flow angle at the SHC exit plane.

Pianko and Wazelt (1983) built up their framework on the earlier works of Wyatt (1955), Tyler (1957), Livesey and Hugh (1966), Traupel (1978) and Dzung (1971). Based on their work, Cumpsty and Horlock (2006) gave a compact review of the different averaging methods with intelligible examples. Moreover, Greitzer et al. (2004) addressed the averaging of generic nonuniform flows. They developed several basic procedures regarding the averaging of the total pressure and demonstrated the capabilities in the case of a two-dimensional channel flow with linearly varying velocity.

It can be shown that four independent integral flow quantities uniquely define a homogeneous one-dimensional flow (Cumpsty and Horlock, 2006). However, this is not possible for a given heterogeneous 3D flow. Performing integral averages will always introduce a certain level of arbitrariness and information loss. This is why it is essential to develop a specific averaging procedure for the desired purpose. For the SHC investigations, the average total pressure, temperature, and velocity components have to be determined at the combustor exit so that the adjacent turbine generates the same power output from the mean uniform flow compared with that generated by the realistic nonuniform flow.

Because the mass and total enthalpy fluxes define together the power conversion of a thermal turbomachine, they are selected as base quantities, which must be conserved by the averaging of the actual nonuniform flow. Furthermore, by neglecting the viscous effects, angular momentum flow and the axial component of the linear momentum flow are chosen as additional invariable quantities. The reason is, that these quantities determine the turbine power output (Euler's turbine equation) and the thrust, respectively, as characteristic output quantities of a gas turbine. In

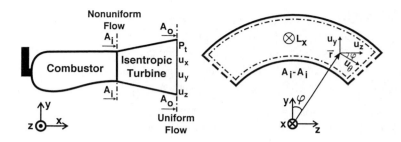

Figure 4.2: The flow conditions and evaluation planes A_i at the inlet and A_o at the outlet of an isentropic turbine (left), and the Cartesian and cylindrical coordinate systems at the turbine inlet plane (right). Dashed lines refer to periodic boundaries, dashed-dotted lines to the contour of the CV at the turbine inlet.

consequence, from those four fluxes, the average velocity components can be derived which represent the whole nonuniform flow at the combustor outlet, i.e., turbine inlet.

The proposed method is based on the following assumptions:

- Isentropic turbine.

- Uniform pressure and velocity profiles at the turbine outlet.

- The working fluid is considered as a perfect gas with a constant mass-averaged specific heat capacity c_p for the calculation of the average total pressure and temperature.

Although the assumption of $c_p = const.$ is not realistic for turbine inlet flow due to the highly varying radial temperature distribution, the effect of considering $c_p = f(T)$ is less than 0.05% as shown by Cumpsty and Horlock (2006). The mean total temperature is simply the mass-averaging of its local value and given by Eq. (4.5) with the preservation of the enthalpy flux and the assumption of perfect gas Eq. (4.3), (4.4). Fig. 4.2 illustrates the coordinate system. The rotation axis of the turbine is parallel to the x-axis, and the turbine inlet A_i and outlet A_o planes are located perpendicular to it.

$$\dot{m} = \int \rho u_x \, dA = \bar{\rho} \bar{u}_x A \tag{4.3}$$

$$\bar{H}_t = \frac{\int H_t \, d\dot{m}}{\dot{m}} \tag{4.4}$$

$$\bar{T}_t = \frac{\int T_t \, d\dot{m}}{\dot{m}} \tag{4.5}$$

For the calculation of the total pressure, the power output of the turbine is considered as the main conserved quantity. Alternatively, other quantities such as the exergy, thrust, or the mass flow through the NGV could be considered, as discussed in detail by Cumpsty and Horlock (2006). The derived average total pressure should be equivalent to the nonuniform total pressure distribution at the turbine inlet plane A_i, in a way, so that the same power output by the turbine is ensured. In an ideal turbine, each particle of the nonuniform flow on plane A_i would experience an isentropic process to reach the turbine outlet plane A_o.

As a consequence of uniform pressure and velocity condition, the field of total pressure is also uniform at the turbine outlet. The power output of the isentropic turbine is then given by

$$\dot{W} = \int c_p(T_{to} - T_{ti})d\dot{m} = \int c_p T_{ti}[(P_{to}/P_{ti})^{\frac{\gamma-1}{\gamma}} - 1]d\dot{m} \tag{4.6}$$

Calculating the average total temperatures by Eq. (4.5) yields

$$\dot{W} = \int c_p \bar{T}_{ti}[(P_{to}/\bar{P}_{ti})^{\frac{\gamma-1}{\gamma}} - 1]d\dot{m} \tag{4.7}$$

Recalling the constant uniform outlet total pressure P_{to} and dividing Eq. (4.7) by Eq. (4.6), the average total pressure at the inlet plane A_i is given by

$$\bar{P}_{ti} = \left[\frac{\int_{A_i} T_t d\dot{m}}{\int_{A_i} T_t/P_t^{\gamma-1/\gamma} d\dot{m}} \right]^{\gamma/\gamma-1} \tag{4.8}$$

Analogously, the expressions for the average fluxes of the axial component of the linear momentum and angular momentum are derived. Reformulating the balance equation of the linear momentum Eq. (2.3) yields

$$\int_{A_o} \rho u_x^2 dA - \int_{A_i} \rho u_x^2 dA = \int_{A_i} P_s dA - \int_{A_o} P_s dA + \Sigma F_{ext} \tag{4.9}$$

which can be written in average one-dimensional form as

$$(\bar{\rho}\bar{u}_x^2 A)_o - (\bar{\rho}\bar{u}_x^2 A)_i = (\bar{P}_s A)_i - (\bar{P}_s A)_o + \bar{F}_{ext} \tag{4.10}$$

The static pressure P_s at A_o is constant as a consequence of the uniform P_t and U. In addition, the distribution of body, friction, and holding forces on the CV through the turbine can be assumed as uniform. Hence, setting Eq. (4.9) equal to Eq. (4.10) gives the average axial momentum flow at the turbine inlet A_i

$$\bar{I}_{xi} = \int_{A_i} (\rho u_x^2 + P_s)dA = (\bar{\rho}\bar{u}_x^2 + \bar{P}_s)A_i \tag{4.11}$$

The average components of angular momentum flow can be derived similarly. For a more straightforward interpretation of angular momentum equations, the radius and circumferential velocity in cylindrical coordinates are introduced as depicted in Fig. 4.2

$$r = \sqrt{y_r^2 + z_r^2} \qquad (4.12)$$

$$tan\varphi = \frac{z_r}{y_r} \qquad (4.13)$$

$$u_\theta = u_z\,cos\varphi - u_y\,sin\varphi \qquad (4.14)$$

Hence, the average components of angular momentum flow are

$$\bar{L}_x = \int_{A_i} \rho u_x \underbrace{\left(y_r u_z - z_r u_y\right)}_{ru_\theta} dA = \dot{m}(\bar{y}_r\bar{u}_z - \bar{z}_r\bar{u}_y) \qquad (4.15)$$

$$\bar{L}_y = \int_{A_i} (\rho u_x^2 + P_s)z_r\,dA = \bar{z}_r\bar{I}_x \qquad (4.16)$$

$$\bar{L}_z = \int_{A_i} -(\rho u_x^2 + P_s)y_r\,dA = -\bar{y}_r\bar{I}_x \qquad (4.17)$$

Recalling the assumptions of the isentropic turbine and the working medium as a perfect gas, the first law of thermodynamics yields

$$\frac{H_t}{H_s} = \frac{T_t}{T_s} = \underbrace{1 + \frac{\gamma-1}{2}M^2}_{C_h} \qquad (4.18)$$

$$\frac{P_t}{P_s} = \left(\frac{T_t}{T_s}\right)^{\gamma/\gamma-1} \qquad (4.19)$$

with

$$M^2 = \frac{\bar{u}_x^2 + \bar{u}_y^2 + \bar{u}_z^2}{\bar{\gamma}R\bar{T}_s} = M_x^2 + M_y^2 + M_z^2 \qquad (4.20)$$

The equations (4.3), (4.11), and (4.15) can be transformed into the following gas dynamic expressions for the mass flow

$$\underbrace{\frac{\dot{m}}{\bar{P}_t A \sqrt{\frac{\gamma}{R\bar{T}_t}}}}_{C_m} C_h^{\frac{(\gamma+1)}{2(\gamma-1)}} = M_x \qquad (4.21)$$

Axial flux of the linear momentum

$$\underbrace{\frac{\bar{I}_x}{\bar{P}_t A} C_h^{\frac{\gamma}{\gamma-1}}}_{C_I} = [1 + \gamma M_x^2] \tag{4.22}$$

Axial flux of angular momentum

$$\underbrace{\frac{\bar{L}_x}{\bar{P}_t A \gamma} C_h^{\frac{\gamma}{\gamma-1}}}_{C_L} = M_x [\bar{y}_r M_z - \bar{z}_r M_y] \tag{4.23}$$

Dividing Eq. (4.22) by Eq. (4.21) yields the relation of the total to static enthalpy as a quadratic function of Mach number in Eq. (4.24), which is known as the Rayleigh relation (Zucker and Biblarz, 2002). Because the supersonic root of Eq. (4.24) violates the second law of thermodynamics, the subsonic root Eq. (4.25) gives the solution for the axial Mach number at the combustor exit.

$$\underbrace{\frac{C_I}{C_m}}_{C_{Im}} \sqrt{C_h} = \frac{1 + \gamma M_x^2}{M_x} \tag{4.24}$$

$$M_x = \frac{C_{Im} \sqrt{C_h} - \sqrt{C_{Im}^2 C_h - 4\gamma}}{2\gamma} \tag{4.25}$$

Similarly, division of Eq. (4.23) by Eq. (4.21) gives Eq. (4.26), which can be solved together with Eq. (4.25) and Eq. (4.20) iteratively to calculate all three components of the velocity field.

$$\underbrace{\frac{C_L}{C_m}}_{C_{Lm}} = \frac{[\bar{y}_r M_z - \bar{z}_r M_y]}{\sqrt{C_h}} \tag{4.26}$$

Because the circumferential velocity component is of main interest in the present work, Eq. (4.14) is applied for calculating the circumferential velocity and Eq. (4.12), Eq. (4.16), and Eq. (4.17) for the mean radius relation required for the circumferential Mach number Eq. (4.27).

$$M_\theta = \frac{C_{Lm} \sqrt{C_h}}{\bar{r}} \tag{4.27}$$

Considering that the Mach numbers are lower than about 0.2 at the combustor exit, the term $\sqrt{C_h} \approx 1.002$ can be neglected. Consequently, the axial and circumferential Mach numbers can be directly calculated by Eq. (4.28) and (4.29)

$$M_x \approx \frac{C_{Im} - \sqrt{C_{Im}^2 - 4\gamma}}{2\gamma} \tag{4.28}$$

$$M_\theta \approx \frac{C_{Lm}}{\bar{r}} \tag{4.29}$$

The average flow angle $\bar{\alpha}$ at the combustor exit plane A_i is then given by

$$tan(\bar{\alpha}) = \frac{M_\theta}{M_x} \tag{4.30}$$

Using Eqs. (4.8) and (4.30), a consistent average total pressure and flow angle value can be derived which represent the real nonuniform flow field at the combustor exit. The calculated average uniform flow delivers the same turbine power output but an incorrect entropy flux compared with the actual nonuniform flow.

In contrast to the method presented here, the conventional exergy (availability) averaging approach, first proposed by Livesey and Hugh (1966) and given by Eq. (4.31), delivers the correct exergy flux but a higher total pressure than the real flow. The reason is that an inhomogeneous temperature distribution represents a potential for work generation (with a reversible heat engine). Accordingly, a uniform flow with the same exergy and enthalpy flux must have a higher pressure than the equivalent nonuniform one. This higher pressure of the uniform one-dimensional flow can then compensate the lost potential to produce work, i.e., exergy, from the averaged-out temperature differences. Cumpsty and Horlock (2006) have discussed this issue in detail based on a simple two-stream model.

$$ln(\bar{P}_{t,exg}) = \frac{1}{\dot{m}} \int_{A_i} ln(P_t)\, d\dot{m} - \frac{\gamma}{(\gamma-1)\dot{m}} \int_{A_i} ln\left(\frac{T_t}{\bar{T}_t}\right) d\dot{m} \tag{4.31}$$

In addition, the linear momentum fluxes, I_y and I_z, as proposed by Livesey and Hugh (1966), yield average velocity components $\bar{u}_y$ and $\bar{u}_z$ given by Eq. (4.32), which are inconsistent with those resulting from angular momentum balance equation in Eq. (4.15).

$$\bar{I}_{yi} = \int_{A_i} \rho\, u_x u_y\, dA = \dot{m}\, \bar{u}_y\, ; \quad \bar{I}_{zi} = \int_{A_i} \rho\, u_x u_z\, dA = \dot{m}\, \bar{u}_z \tag{4.32}$$

The methods introduced in this chapter are embedded in automated post-processing scripts, which are applied to the results of a numerical parametric study of the SHC concept. Hence, the reacting helical flow could be evaluated globally, quantitatively, and consistently within the unconventional flame tubes.

In the next chapter, the basic rules for the design and scaling of a generic SHC are introduced.

5 Combustor Scaling and Similarity Analysis

After establishing a framework for numerical investigations, first in this section, some elementary considerations are discussed regarding the scaling of a novel combustor and the associated similarity laws. On that basis, for the particular case of the SHC the consequences of the staggered arrangement of the burners on the reacting flow in the flame tube are elaborated. Accordingly, some critical design parameters are determined and their limits are discussed. As a result, suitable approaches for scaling the flame tube and burners of the SHC are proposed.

Scaling of combustion systems involves the consideration of aerodynamics, heat transfer, and combustion in the vicinity of the burners as well as in the whole combustor. Spalding et al. (1963) and Beér and Chigier (1972) summarized and listed a large number of dimensionless groups derived from the balance equations of mass, momentum, and energy. Requesting similarity in terms of the fundamental dimensionless numbers, e.g., Reynolds, Mach, and Damköhler, leads to conflicting requirements. Hence, the central problem of modeling and scaling of combustion systems is to discern which degree of model abstraction and simplification is appropriate and consequently which scaling laws are to be fulfilled or neglected. The term "partial modeling" was introduced by Spalding et al. (1963) for this approach. In this context, different scaling techniques are in use for combustion engineering, which have been reviewed and summarized by Gupta and Lilley (1985) and Weber (1996). The relevant burner and flame tube scaling laws for a parametric study of the SHC are introduced below. The effects of flame tube and burner scaling on the overall similarity of swirling flows and heat transfer to the liners are discussed within the evaluation of the parametric study in Sec. 6.1.

In combustion engineering, the term "scaling analysis" often also refers to the investigation of a certain combustor geometry with respect to different operating pressures. The following scaling analysis, however, is focused on full load conditions of an aircraft engine with the corresponding constant operating pressure and thermal power output.

5.1 Flame Tube Scaling

As discussed in Sec. 2.2.3, gas turbine combustion takes place mainly in regimes characterized by very large Damköhler numbers $Da = t_{\tau}/t_f \gg 1$. That means the chemical reactions are typically much faster than the turbulent mixing processes $t_f \ll t_{\tau}$. Thus, t_{τ} is the main controlling parameter and affect the combustion performance and pollutant formations directly. Thus, it has to be examined how the scaling changes mixing phenomena in the combustion system. Recalling the turbulent scales from Sec. 2.1.3, the integral eddies control the macromixing while the micromixing takes place at Kolmogorov scales (Bockhorn et al., 2009).

The two most frequently applied scaling methods for industrial combustion systems are the Constant Velocity Scaling (CVS) and Constant Residence Time Scaling (CRTS) (Weber, 1996). For the flame tube, the CRTS is applied, which maintains the macromixing constant, whereas the micromixing may be altered. The objective is to ensure an overall preserved combustion performance and emission characteristics.

It is known that pollutants, and in particular NOx emissions, correlate with the combustor global residence time (Driscoll et al., 1992, Peters and Donnerhack, 1981), which is maintained as constant by the CRTS method. However, Hsieh et al. (1998) showed for swirl-stabilized combustion, that the local residence time in the flame recirculation zones is a critical parameter for pollutant emissions as well. As shown in Sec. 5.2, the effect of flame residence time will be indirectly considered by the scaling of burners because the volume of the IRZ is directly dependent on the burner diameter.

The characteristic geometric length of the combustor flow $\mathcal{L}$, typically the height of the flame tube, determines the turbulent integral length scale $\ell_{\mathcal{I}}$. $\mathcal{L}$ can be estimated as at least one order of magnitude larger than $\ell_{\mathcal{I}}$ (Peters, 2000). Analogously, the integral time scale $t_{\mathcal{I}}$ can be estimated with the mean streamwise velocity U in the combustor as

$$t_{\mathcal{I}} \sim \mathcal{L}/U \tag{5.1}$$

Furthermore, the ratio $\mathcal{L}/U$ can be estimated by the combustor global residence time t_{res} so that

$$t_{\mathcal{I}} \sim \mathcal{L}/U \sim t_{res} \sim V/\dot{V} \tag{5.2}$$

Following Eq. (5.2), a constant thermal power ($\dot{W}$) as well as volumetric fluid flow rate ($\dot{V}$) into the SHC are to be considered. For maintaining the integral time scale $t_{\mathcal{I}}$ constant, in a first-order approximation, the volume of SHC must be kept the same as the Reference Conventional Combustor (RCC)

$$V_{shc,crts} = V_{rcc} \quad \xrightarrow[\dot{W}_{shc} = \dot{W}_{rcc}]{\dot{V}_{shc} = \dot{V}_{rcc}} \quad t_{res,shc} = t_{res,rcc} \longrightarrow \quad t_{\mathcal{I},shc} \approx t_{\mathcal{I},rcc} \tag{5.3}$$

The micromixing effects on the flame characteristics can be rated by the Karlovitz number as introduced by Eq. (2.22). Assuming a sufficiently large distance to the broken reaction zone, see Sec. 2.2.3, the chemical length scale t_f can be considered as invariant regarding the turbulence. The effect of combustor scaling on the variation of the Karlovitz number would be then only dependent on the behavior of the Kolmogorov length scales t_κ. This can be estimated in turn by the relation of the integral and Kolmogorov length scales to the turbulent Reynolds number as given in Eq. (2.20). The flow Reynolds number, on the other hand, scales reciprocally with the characteristic combustor length $\mathcal{L}$, which results from CRTS and the condition of constant volumetric flow rate

$$Re \sim \mathcal{L}U \quad \xrightarrow[U \sim \mathcal{L}^{-2}]{\dot{V} = const.} \quad Re_{crts} \sim \mathcal{L}^{-1} \tag{5.4}$$

Accordingly, the similarity of ℓ_κ with $\mathcal{L}$ can be estimated as

$$\ell_{\mathcal{I}}/\ell_\kappa \sim Re^{3/4} \quad \xrightarrow[Re_{crts} \sim \mathcal{L}^{-1}]{\ell_{\mathcal{I}} \sim \mathcal{L}} \quad \ell_{\kappa,crts} \sim \mathcal{L}^{7/4} \tag{5.5}$$

The Karlovitz number would be then characterized by

$$Ka_{crts} \sim \ell_\kappa^{-2} \sim \mathcal{L}^{-7/2} \qquad (5.6)$$

Thinking about a 50% upscaling of the height of the flame tube, e.g., for a double annular configuration, the overall Karlovitz number would be reduced by a factor of around 4. For realistic operating points of lean gas turbine combustion ($p > 30$ bar), there is a wide margin of about three orders of magnitudes between the boundaries of the broken and thin reaction zones $Ka_\delta \approx 0.001 Ka$. Taking a Karlovitz number of $\mathcal{O}(1)$ into account, it can be assumed that for either up- and downscaling of flame tube dimensions of about 50%, the combustion still takes place in the thin reaction zone regime.

The preceding analysis shows that, to the first order of accuracy, the CRTS ensures the similarities of the flow field and combustion regime in the flame tube. Thus, for all SHC models investigated in the present study, a constant volume of the flame tube and the same total mass flow through the combustor are considered.

The structure of turbulent flames, however, is also significantly influenced by local flow features, which to a great extent, are determined by the burner aerodynamics. The effect of burner scaling on combustion characteristics is discussed in the following section.

5.2   Burner Scaling

The CVS method is applied for the scaling of the burners. With this method, the bulk velocity in the burners is maintained as a constant. Thus, a constant pressure drop and a preserved flame–vortex interaction are expected.

The number of burners N_b in the SHC can be varied by changing the number of flame tube segments N_{seg}, or by switching to a double annular design. With CVS, the diameter of the burners d_{cvs} is calculated by

$$d_{shc,cvs} = d_{rcc}(N_{b,rcc}/N_{b,shc})^{1/2} \qquad (5.7)$$

Thereby, with $\ell_\mathcal{I} \sim d$ the integral scales as well as the Reynolds number can be estimated as inversely proportional to the square root of the number of burners

$$\ell_{\mathcal{I},cvs} \sim t_{\mathcal{I},cvs} \sim Re_{cvs} \sim d_{cvs} \sim N_b^{-1/2} \qquad (5.8)$$

Analogous to Eq. (5.5), the Kolmogorov length scale can be estimated by

$$\ell_\mathcal{I}/\ell_\kappa = Re^{3/4} \quad \xrightarrow[Re_{cvs} \sim N_b^{-1/2}]{\ell_{\mathcal{I},cvs} \sim N_b^{-1/2}} \quad \ell_{\kappa,cvs} \sim N_b^{-1/8} \qquad (5.9)$$

Again, with the assumption of constant chemical time and length scales, the Damköhler and Karlovitz numbers can be estimated with

$$Da_{cvs} \sim t_{\mathcal{I},cvs} \sim N_b^{-1/2} \quad , \quad Ka_{cvs} \sim \ell_{\mathcal{K},cvs}^{-2} \sim N_b^{1/4} \tag{5.10}$$

Accordingly, by doubling the number of burners based on the CVS method, the corresponding flow exhibits a reduction of the local Damköhler number by about 30%, as well as an increase of the local Karlovitz number by about 20%. This can also be explained, as mentioned in the previous section, by the reduced local flame residence time due to smaller IRZ as reported by Hsieh et al. (1998). The effect of CVS on the NOx emissions was investigated by Smart et al. (1992) and Weber and Breussin (1998). The authors reported higher NOx emissions, which were attributed to high thermal-NO formation rates due to temperature peaks corresponding to the reduced turbulent mixing.

The effect of reduced mixing performance of the downscaled burners in a double annular arrangement, however, is expected to be compensated by the additional turbulent exchange between the radially adjacent flames. Furthermore, due to the CRTS of the flame tube, the overall combustor residence time is maintained constant, which implies a longer time for the post-flame turbulent mixing in the case of the compact swirling flames. Besides, the flow Reynolds numbers, also for the downscaled burner diameters considered in the SHC parametric study, are always of $\mathcal{O}(5)$. Thus, the large Reynolds number limit would not be violated for any configuration of the parametric study.

Similar to the analysis of micromixing in CRTS, the changes in Karlovitz numbers within the operating parameters ($\Delta Ka_{cvs} = \pm 20\%$) would not change the combustion regime. Moreover, it should be noted that the geometrical variations in the burner and flame tube are acting counterbalancing regarding the Karlovitz number. So, increasing the height of the flame tube, as a requirement for a double annular configuration, would reduce the overall Karlovitz number in the flame tube whereas the smaller burner diameters lead to higher local Karlovitz numbers.

Furthermore, as discussed in Sec. 1.2.2, for the scaling of the burners, it is essential to preserve the overall features of the swirling flow. Thus, at least the burner confinement ratio, the type of inlet vortex and the inlet swirl number must be maintained constant. The former one is considered as a control parameter in the SHC parametric study. The similarity of the vortex type and the swirl number must be sustained by the appropriate aerodynamic design of the scaled swirlers. This is achieved in the present study by adapting the inlet swirl velocity profile of the burners. Also the flame size in terms of its axial extent, might affect the combustor exit flow pattern, which is a decisive parameter in current SHC study. Therefore, the flow pattern similarity is discussed additionally in the SHC parametric study in Sec. 6.1.1.

At this point, the necessary framework for a fundamental study and optimization of the SHC concept is completed. It features a robust and experimentally validated numerical setup, the postprocessing tools for a global and physically consistent assessment of the combustor flow, and finally design and scaling rules based on the similarity considerations regarding the gas turbine combustors. In the following chapter, the results of the parametric study and possible strategies to control the flow and optimize the SHC concept are introduced.

6 The Short Helical Combustor Concept

In this chapter, a generic SHC model is introduced and analyzed through a parametric study. The similarity laws discussed in Chap. 5 are taken into account to derive scaling rules for the geometry of the flame tube and the burners. Thereafter, with the numerical tools introduced in Chap. 2 the nonreacting and reacting flow regimes of selected design variants are investigated.

The flow structure of different design configurations is studied by a kinematic assessment of the velocity fields. The underlying physics are elucidated by analysis of the flow dynamics via the integral balance of angular momentum based on the methods introduced in Chap. 4.

Finally, with the insight gained from those investigations, a generic design improvement is proposed to address the tremendous loss of initial angular momentum flow, as well as the inhomogeneous flow and temperature field at the outlet of the SHC. As a result, the proposed SHC features high angular momentum flow as well as uniform flow angle and temperature fields at the exit plane.

6.1 Similarity Study

As a first step of the similarity study, a generic conventional annular combustor is proposed, which is referred to as the RCC. Correspondingly, an SHC variant is derived from the RCC. The relevant dimensions of both combustors are illustrated in Fig. 6.1.

The RCC is based on the combustor of a Rolls-Royce BR725 turbofan aircraft engine. It features swirl-stabilized lean combustion with 14 burners in a single annular arrangement. The annular flame tube is unwrapped and modeled by prismatic segments. Furthermore, the mean radius of the flame tube is constant from the dome to the outlet. The SHC design features a tilting angle β between the burner axis and turbine shaft. Accordingly, the subscript β of geometric parameters will indicate the SHC type with the associated tilting angle. In all SHC variants to be studied, the flame tube outlet height and the R_{mean} are identical to the RCC and maintained constant.

As discussed in Sec. 5.1, the SHC flame tubes are scaled based on the CRTS method. Maintaining the same volume of the SHC flame tubes as the RCC, a general scaling law can be derived on the basis of Eq. (5.3). Accordingly, the number of flame tube segments and the tilting angle of the burner axis determine the pitch of the staggered segments

$$V_{\beta,crts} = L_{seg,\beta} \, H_{seg,\beta} \, P_{seg,\beta} \, N_{seg,\beta} = V_{rcc} \tag{6.1}$$

$$P_{seg,\beta} = P_{seg,rcc} \, \cos\beta \, \frac{N_{seg,rcc}}{N_{seg,\beta}} \tag{6.2}$$

The reduction of the combustor volume due to the smaller pitch must be compensated by adapting

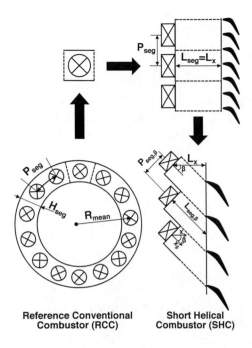

Figure 6.1: Development of a generic *planar* SHC model from a reference conventional *annular* combustor. Dashed lines refer to the boundaries of the individual flame tube segments.

the combustor length and/or height. Eq. (6.1) and Eq. (6.2) yield

$$L_{seg,\beta} \, H_{seg,\beta} = \frac{L_{seg,rcc} \, H_{seg,rcc}}{\cos \beta} \tag{6.3}$$

According to Eq. (6.3), there are various scaling approaches possible. A general scaling law with linear (Λ, Ξ) and exponential (λ, ξ) weighting factors is given by Eq. (6.4, 6.5).

$$L_{seg,\beta} = \frac{\Lambda}{\cos \beta^{\lambda}} L_{seg,rcc} \quad , \quad H_{seg,\beta} = \frac{\Xi}{\cos \beta^{\xi}} H_{seg,rcc} \tag{6.4}$$

$$\lambda + \xi = 1 \quad , \quad \Lambda \, \Xi = 1 \tag{6.5}$$

A reasonable and straightforward approach is to choose $\lambda = 0$ and $\Lambda = 1$ so that the length of the flame tube segment is identical to that of the RCC, as depicted in Fig. 6.1. This scaling law will be denoted L0 and is given by

$$L_{seg,\beta} = L_{seg,rcc} \quad \longrightarrow \quad H_{seg,\beta} = \frac{H_{seg,rcc}}{\cos \beta} \tag{6.6}$$

Applying the L0 scaling approach, a parametric study is conducted, which involves variation of the tilting angle, the number of segments, as well as the arrangements, single annular (SA) and double annular (DA).

The burner confinement ratio, as mentioned in Sec. 5.2, can strongly affect the structure of the swirling flow. It also represents the minimum distance between flames and confining walls and, thus, is crucial regarding flame quenching and wall heat transfer.

Therefore, the burner confinement ratio is defined as the main control parameter:

$$\Delta_\beta = \frac{d_{max}}{d_\beta} \tag{6.7}$$

d_{max} in Eq. (6.7) corresponds to the maximum burner diameter installable in a segment as illustrated in Fig. 6.3.

6.1.1 Flow Pattern Similarity

In Fig. 6.2, the burner confinement ratio Δ of selected L0 scaling results are plotted as a function of the parameters SA, DA, number of segments N_{seg}, and tilting angle β. It should be noted that the curves of L0-7SA and L0-14DA coincide.

As an example of the applied notation, L0-14DA45 refers to an SHC type with L0 scaling, 14 segments, DA configuration, i.e 28 burners, and a tilting angle $\beta = 45°$. In Fig. 6.3, the characteristic variants are compared via true-to-scale pictograms.

For different tilting angles β, the burner confinement ratio in the SHC can be kept the same as for the RCC. For that, the number of flame tube segments must be reduced when increasing β. Thereby, fewer burners can be compensated either by upscaling or using multiple burners per segments, e.g., by a DA arrangement. With the same configuration as RCC, i.e., SA and 14 original burners, a relatively low tilting angle of $\beta = 25°$ can be realized with L0-14SA25. The original RCC burners can also be used in the L0-7DA60 configuration. It features 14 burners in a DA arrangement with a tilting angle of $\beta = 60°$. However, a feasible tilting of the burners must be matched with the air discharging the last compressor rotor blades. It typically has a mean flow angle of about 45°. Thus, a tilting of smaller than 45° would not utilize the whole potential of the SHC concept, whereas a higher tilting would be associated with a complicated design of a vaned prediffuser with surge hazard, as well as huge sidewalls.

Consequently, the most suitable SHC embodiments would be those with a tilting of $\beta = 45°$, i.e., the L0-7SA45 and L0-14DA45 types. However, for these types, the burners must still be scaled based on the CVS method. As mentioned previously, the scaling has an impact on the size of swirling flames, which in turn may affect the structure of the primary combustion zone and the combustor exit flow pattern.

The effect of burner scaling on the flow pattern is examined by the parameter

$$\Pi_\beta = L_{seg,\beta}/x_{rb} \quad \text{with} \quad x_{rb} \sim d_\beta \tag{6.8}$$

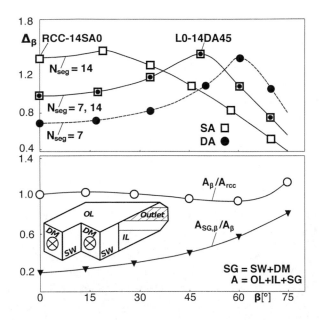

Figure 6.2: The parametric study using the L0 scaling law. Burner confinement ratio Δ (top) and flame tube surface ratios (bottom) as functions of the tilting angle β. OL: outer liner, IL: inner liner, SW: sidewall, DM: dome, SG: staggered wall.

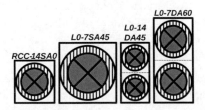

Figure 6.3: Schematic of RCC and selected SHC types. Gray circles correspond to the actual burner diameter d_β and hatched one to d_{max}.

where x_{rb} is the axial extent of the recirculation bubble, and can be estimated as proportional to the burner diameter. The other major parameters determining the type of vortex break down and the size of recirculation bubbles are maintained constant. These are the burner confinement ratio and the swirl number, which are investigated in detail for a DA–SHC configuration by Holz (2015). The distance between the primary zone and the combustor exit is a measure of the postcombustion mixing, and thus affects the average flow and temperature patterns at the exit. Hence, to the first order of accuracy, the parameter Π_β will reflect the SHC exit flow pattern based on the size of the swirl flames.

With Eq. (5.7), the flame size is estimated to be $x_{rb} \sim N_{b,\beta}^{-1/2}$. Hence, the proposed pattern factor in normalized form is given by

$$\Pi_{\beta,Nb} = \frac{x_{rb,rcc}}{x_{rb,\beta}} \frac{L_{seg,\beta}}{L_{seg,rcc}} \frac{L0,CVS}{} \left(\frac{N_{b,\beta}}{N_{b,rcc}}\right)^{1/2} \tag{6.9}$$

where N_b refers to the number of burners. For the L0-7SA45 type the pattern factor is $\Pi_{45,7} = 0.7$. Reducing the number of burners and scaling them up leads to large swirl flames, and consequently to poor circumferential uniformity at the combustor exit. This was confirmed by CFD simulations of the L0-7SA45 SHC type and reported by Wilhelm (2013).

In contrast to seven large flames of L0-7SA45 or fourteen moderate flames of RCC, the L0-14DA45 features twenty-eight compact swirling flames. With $\Pi_{45,28} = 1.4$, the relative distance to exit can be estimated to be 40% longer than the RCC. Hence, more uniform exit flow and temperature patterns might be attained in an L0-14DA45 design.

The reduction of burner diameters, as discussed in Sec. 5.2, is expected to have counteracting effects regarding the pollutant emissions. The local turbulent mixing and flow residence time in the compact swirl flames are decreased on the one hand, whereas on the other hand the post flame time and the lateral turbulent exchange between the radially adjacent flames are increased.

The nonreacting flow simulations of the SA and the similar DA–SHC types showed in general, as reported in Ariatabar et al. (2014), more uniform flow fields in the primary combustion zone as well as at the flame tube exit of the DA configurations. Accordingly, a positive net effect on the turbulent mixing of the flow can be attributed to the DA–SHC designs.

A general statement regarding the improvement of the emission characteristics of such a DA–SHC design, however, should be treated with caution. For that, in an advanced numerical and experimental study of the reacting flow field in SHC, the nonlinear effects of local aerothermal inhomogeneities on the pollutant formation must be investigated.

Regarding the combustion stability in the L0-14DA45, the additional turbulent exchange of heat and momentum, which takes place perpendicularly to the main flow direction, implies a supporting stabilizing effect between the radially adjacent flames. Furthermore, the vortex breakdown instabilities of the compact swirling flames are less susceptible to aerothermal oscillations compared with those of large flames. On the other hand, as discussed in Sec. 5.2, the local Damköhler number in the individual flames is expected to be reduced by around 30%, which might trigger the quenching of the flames, and thus impair the stability characteristics of the L0-14DA4.

Besides the geometry parameters like the flame tube dimensions, the design of the dome, and the exit part as well as the arrangement of the burners, also the combustor operation conditions are of crucial importance to the stability characteristics of a combustion system. While a detailed stability analysis of SHC is beyond the scope of the current study, a comprehensive review of stability characteristics of lean-premixed swirl-stabilized combustion can be found in Huang and Yang (2009).

In addition to the flow similarity, a further important issue in development of the SHC is the cooling of the additional sidewalls and the associated heat transfer similarity. This issue is addressed in the following section.

6.1.2   Liner Heat Transfer Similarity

A characteristic parameter reflecting the heat transfer to the combustor liner is the ratio of the total surface area to the combustor volume. Because the volume of all variants is identical, the parameter A_β/A_{rcc} is used for the comparison to RCC. It can be seen in Fig. 6.2 that the cooling surfaces of the L0-14DA45 and L0-7SA60 types are reduced by approx. 5% and 10%, respectively. Thus, the overall heat transferred to the flame tube walls of these models can be assumed as nearly unaltered.

The total surface of the thermal heavily loaded staggered wall $A_{sg,\beta}$, however, increases with the tilting angle. In the case of the L0-14DA45, $A_{sg,\beta}$ takes 40% of the total SHC surface A_β. The ratio $A_{sg,\beta}/A_\beta$ is approx. 60% for the L0-7SA60 type. It should be noted that in the case of RCC, the dome provides only 20% of the total combustor surface. It is most likely that a considerable part of the cooling air, which might be saved due to the reduction of NGV, must be invested in cooling the sidewalls.

Summary of the Parametric Study

A general scaling law for the SHC has been derived based on a simple generic combustor model. On that account, a parametric study has been undertaken. The results showed that the SHC type L0-14DA45 could satisfy the relevant geometric, aerodynamic, heat transfer and combustion similarity parameters. This type exhibits up to 30% shorter axial length at a tilting angle of $\beta = 45°$.

On the other hand, it must be considered that for a DA design, the prediffuser must be adapted to provide uniform airflow for the radially arranged burners. Moreover, doubling the number of burners requires a more complex fuel supply system as well as manufacturing and mounting efforts.

Numerical investigations of other SHC configurations like SA designs L0-14SA25 and L0-7SA45 were carried out by Ariatabar et al. (2015a,b, 2016). Those studies confirm the result of the present scaling and similarity analysis, that the SA design is not suitable for the SHC, at least not for an axial configuration.

For radial combustor designs as typically utilized in compact helicopter engines, however, an SA–SHC with high tilting angle of burners could be relevant. The reason is that in those cases, the combustor dome is located at a higher mean radius than the NGV. Therefore, in contrast to the axial combustors, the pitch-to-height ratio of segments is typically higher than one; as an example, Philip et al. (2015) and Staffelbach et al. (2009) reported $P_{seg}/H_{seg} \approx 1.5$. Hence, a high tilting of segments is still possible without it being necessary to reduce the number and increase the diameter of the burners. Such a centripetal SHC is proposed by Negulescu (2014) and shown in Fig. 1.2. This design of the SHC is, however, out of the scope of the present study.

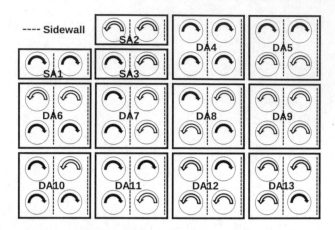

Figure 6.4: Possible swirl rotational directions of SA and DA–SHC.

Based on the present scaling and similarity study, the focus of the following numerical analyses is laid on the L0-14DA45 types.

6.2 Numerical Analysis

The scaling and similarity analysis of the SHC concept showed that a configuration with a DA arrangement and a burner tilting angle of $\beta = 45°$ is superior. It exhibits an optimum with respect to the most critical similarity issues and exploiting the potential of burner tilting to shorten the length of the combustor and reduce the number of the NGV.

It is, however, still not clear how the asymmetrically bounded swirling flows would behave in such an SHC configuration. Furthermore, it is unknown what would be the resulting mean flow angle at the outlet and in front of the NGV, and if the rotational direction of the adjacent swirling flows relative to each other would be a critical parameter regarding the flow structure.

Hence, in the first step in this section, the interactions of adjacent swirls are investigated. The flow kinematics in the primary combustion zone, as well as the exit flow pattern, are analyzed based on numerical simulations. Fig. 6.4 shows the schematics of all pairing possibilities of the adjacent swirl rotational directions, where each of them is illustrated by two segments of the flame tube with periodic boundaries.

The swirl configurations 1–3 refer to the SA arrangement and are shown for the sake of completeness. As discussed in Sec. 6.1.1 because of the very large flames and nonuniform outlet flow patterns, these configurations are not considered for further investigation and optimizations. The DA configurations 4–9 are the main subjects of the numerical analyses in the following sections.

The configurations 10–13 illustrate the possibilities, where the outer rows are corotating and the inner rows counterrotating, and vice versa. They can be interesting with respect to the staging of the burners, where for example the inner row is designed as the pilot and the outer row as the main stages. In that scenario, the different swirl configuration of the inner and outer rows might be advantageous regarding their different functionality and aerothermal requisites. For an investigation of the principal flow structure of this category, the DA10 configuration is selected for further numerical simulation.

Before introducing the simulation results, the applied numerical setup, as well as the initial and boundary conditions, are briefly reviewed in the next section.

6.2.1   Numerical Setup

The numerical setups are principally in accordance with those introduced in Sec. 3.2 for the Preccinsta model combustor. The computational domain has approx. 1 million elements per flame tube segment. The outlet is extended with a buffer channel to ensure the quality of the solution on the SHC exit plane. Furthermore, a nonreflecting pressure boundary condition is applied at the outlet to prevent wave reflections and to maintain better numerical stability.

The applied boundary and initial conditions are based on the take-off operating conditions of a jet engine of a regional civil aircraft. The pressure is $p = 31$bar and the inlet temperature $T_{in} = 820$K. The turbulent intensity at the inlet is set to be 5% of bulk velocity. The walls are treated as adiabatic with standard wall functions. The average dimensionless wall distances on the flame tube were approximately $y^+ = 100$ for the cold and $y^+ = 250$ for the reacting flow simulations.

The total air mass flow through the combustor is $\dot{m}_{air} = 42kg/s$. 70% of the air is considered to be passed through the burners, which is typical for a lean combustion system. The remaining 30% would be used primarily for cooling but is not considered in this study. The reason is to focus on the interaction of staggered swirling flames. However, the interaction of swirling flames with a possible cooling film on the sidewall might be a critical issue regarding static but also dynamic combustion stabilities, which must be considered in other advanced SHC studies.

To establish realistic flow patterns and intensive shear layers, the inlet is modeled as a two-stage coaxial swirler, with boundary conditions similar to the burner used by Bärow et al. (2013) and Gepperth et al. (2014). It consists of a tube and a coaxial annulus. The swirl number of the flow through the annulus is $S_{annulus} = 1$, through the tube $S_{tube} = 0.5$. The mass flow averaged swirl number of the burner is $S_b = (\dot{m}_{tube} S_{tube} + \dot{m}_{annulus} S_{annulus})/(\dot{m}_{tube} + \dot{m}_{annulus}) = 0.83$ with $\dot{m}_{annulus}/\dot{m}_{tube} = 2$.

Klatt (2012) and Trapp (2014) investigated the modeling of the reference radial swirler by imposing vortex velocity profiles at the inlet of straight cylindrical tubes. Their parametric studies involved Rankine-vortex profiles with an increasing portion of the forced-vortex. In both studies, the reproducing of the flow field of the reference swirler was performed best by Rankine-vortex profiles featuring a forced-vortex proportion of about 50%. However, the corresponding simulations were prone to numerical instabilities due to the higher maximum values and higher radial gradients of the tangential velocities. These implied smaller time steps, and thus higher total computational costs. On the other hand, the simulations with the pure forced-vortex profiles feature technically also a free-vortex portion of about 10%, following the no-slip condition at

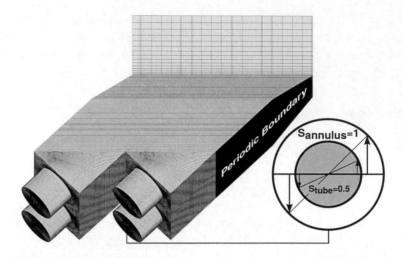

Figure 6.5: The CFD domain of the *L0-14DA45* SHC type with an illustration of the imposed swirl profiles at the inlet.

the walls. Moreover, the forced-vortex profiles provide a qualitatively good reproduction of the swirler flow field with about 20% lower numerical cost and more robust computational procedure. Therefore, for the SHC parametric studies the reference radial swirler was modeled by imposing forced-vortex velocity profiles at the inlet as shown in Fig. 6.5.

Total simulation time corresponds to 12 flow cycles through the combustor. The calculation of the time-averaged flow quantities is started after three establishing flow cycles.

6.2.2   The SHC Flow Kinematics

The kinematic description of the flow does not consider how the motion is brought about by disregarding the forces that cause the fluid motion. Hence, in the kinematics framework, the balance equations and conservation laws are not dealt with. An introduction to kinematic flow analysis can be found in Schobeiri (2010). In this context, the interaction of swirling flows is investigated for selected SHC types based on the contour plots of velocity and flow angle on a system of evaluation planes as shown in Fig. 6.6.

The results of kinematic studies can be used as a groundwork that is necessary for describing the dynamics of the fluid, which will be addressed in the subsequent section.

The DA4,5,6,9 types, as shown in Fig. 6.4, can be modeled using one flame tube segment with periodic boundaries. Similarly, the DA7,8 types can be modeled with only one segment using symmetric boundaries. The DA10–13 types, however, need to be modeled using at least two segments with periodic boundaries. Hence, to ensure comparability of the calculations, all variants are simulated using two segments and periodic boundary conditions. Hereafter, the results of the DA4,6,7,10 types will be addressed. They exhibited characteristic flow features

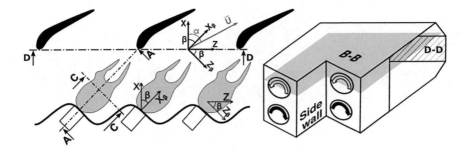

Figure 6.6: Schematic illustration of the evaluation planes A–A, B–B, C–C, and D–D of the SHC. The C–C planes are located normal to the burner axis at the distance $x_\beta/d_\beta = 0.7$

and extreme average values of the exit flow angle. The flow patterns of other types were similar to these and can be deduced based on their swirl configurations and the physical phenomena, which will be discussed below.

Nonreacting Flow

The structure of the flow in the vicinity of the burners is shown in Fig. 6.7. The contour plots show the velocity component $U_{x,\beta}$ parallel to the tilted burner axis at the C–C planes. The interaction of radially adjacent swirls leads to the cross-sectional stagnation of the swirling jets. They are consequently accelerated parallel to the burner axis. This can be recognized by the high-velocity regions, which are marked in Fig. 6.7 by dash-dotted lines. Depending on the swirl rotation direction of radially adjacent burners, the swirling jets are located horizontally for the counterrotating (DA6,10) or inclined for the corotating (DA4,7) types.

The inclined swirling jets in the case of the DA4,7 types lead to deformation of the recirculation bubbles, which might incur a high thermal load on the sidewall or trigger flame quenching.

In contrast, the swirl cores of the DA6 type exhibit radial symmetry. At the middle height, the flow is uniform and directed toward the adjacent segment, as schematically indicated by the arrow in Fig. 6.7. This can be advantageous regarding the advection of the combustion products to the neighboring flames, thus, promoting a "neighbor-piloting" effect with enhanced ignition and flame stability.

The DA10 type exhibits a combination of the previously discussed phenomena. The outer swirls are counterrotating, thus, similar to those in the DA7 type, they penetrate in the adjacent segment. The mixing with the neighbor counterrotating swirl core leads to distorted asymmetric recirculation bubbles.

Regarding the overall uniformity and symmetry of the flow, the DA6 type (radially counterrotating, circumferentially corotating swirls) is the best and the DA7 type (radially corotating, circumferentially counterrotating swirls) the worst swirl configuration. The flow fields of these SHC types are presented via the circumferential B–B planes in Fig. 6.8, the longitudinal A–A planes in Fig. 6.9, as well as the exit planes D–D in Fig. 6.11.

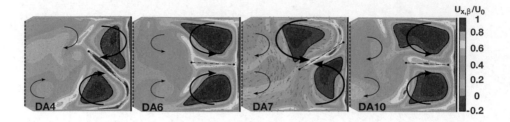

Figure 6.7: Nonreacting flow. Contour plots of the time-averaged velocity field with isolines of $U_{x,\beta} = 0$ on the C–C planes. Arrows indicate the swirl cores and dash-dotted lines highlight the cross-sectional stagnation of the swirling jets. The solid lines on the right side of the plots mark the sidewalls, and the dashed lines on the left sides mark the periodic boundaries.

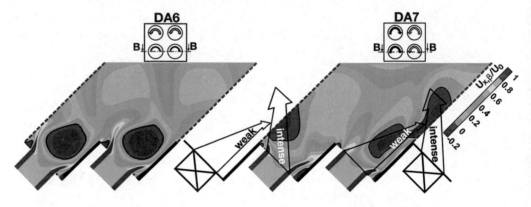

Figure 6.8: Nonreacting flow. Contour plots of the time-averaged velocity field with isolines of $U_{x,\beta} = 0$ on the B–B planes. The size of arrows represents the intensity of the swirl cores.

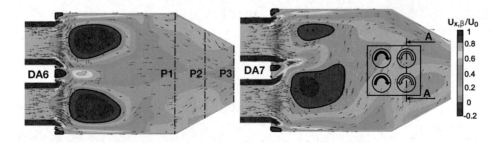

Figure 6.9: Nonreacting flow. Contour plots of the time-averaged velocity field with isolines of $U_{x,\beta} = 0$ on the A–A planes. The dash-dotted lines P1–P3 indicate the evaluation planes parallel to the D–D plane, for calculation of the mean flow angle α and its standard deviation σ_α.

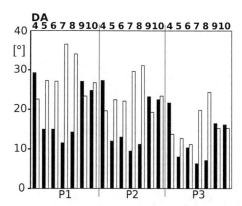

Figure 6.10: Nonreacting flow. Bar diagram shows the average flow angle (filled bars) and its standard deviation for the DA4-10 types. The P1–3 evaluation planes are defined in Fig. 6.9.

As shown in Fig. 6.8, the swirl cores of the DA7 are deflected pairwise toward each other. This happens due to the inclined placement of the radially adjacent swirling jets as marked in Fig. 6.7 by the dashed-dotted lines. Because of the staggered arrangement, the neighbor cores have unequal intensities at the interacting points on the segment boundaries. Hence, the swirl core with higher intensity either pulls the weaker adjacent core into its own segment or pushes it off and forces itself into the neighboring segment. The resulting flow pattern is extremely inhomogeneous, and the recirculation zones are extended almost into the convergent exit part of the flame tube (see the DA7 in Fig. 6.8 and 6.9).

The DA6 type shows, in contrast to other configurations, an almost symmetric flow structure both circumferentially and radially. The recirculation bubbles are relatively compact and located in the vicinity of the burners. Thus, the interactions of the swirl cores with each other and with the sidewalls are minimized.

At the SHC exit, on the other hand, the flow must exhibit a turbine-conform pattern and simultaneously a large mean flow angle. The mean flow angle α and its standard deviation σ_α were evaluated on the P1–P3 planes as shown in Fig. 6.9. The averaging is performed with the physically consistent method proposed in Sec. 4.2. The results are shown as a bar diagram in Fig. 6.10. The flow angle is decreasing continuously as the flow is passing the convergent exit part of the flame tube. This can be explained by the continuity and balance equation for angular momentum. The contraction of the flow cross section leads to the acceleration of the fluid normal to the combustor exit plane. Thus, U_x is increased. At the same time, due to the balance equation of angular momentum, $RU_z = const.$, the circumferential velocity component U_z stays constant on average. The latter holds because the mean radius of the flame tube, unlike the majority of gas turbine combustors, in the present SHC models is constant. As a result, the flow is deflected toward the axial direction, and the mean flow angle $\bar{\alpha}$ is reduced.

At the same time, the flow field becomes more uniform, which is quantified by lower standard deviations of the flow angle. The DA4 exhibits the highest average exit flow angle of $\bar{\alpha} \approx 20°$.

The other extreme is DA7 with $\bar{\alpha} \approx 5°$ and $\sigma_\alpha \approx 20°$, which is a result of the massively deformed recirculation bubbles in the primary zones.

The flow fields at the combustor exit planes are shown in Fig. 6.11. The DA4 exhibits a large region of highly angled flow next to the outer liner. However, the radial distribution of the angled flow is asymmetric, which is suboptimal for turbine matching. The DA6 exhibits a lower mean flow angle than DA4. However, it might be advantageous for the NGV aerodynamics and heat transfer design because the flow pattern is radially symmetric with large flow angles at the middle height and low at the endwalls. The exit flow patterns of the DA7 and DA10 types are highly nonuniform both circumferentially and radially. For the DA7 type, the alternating positive and negative flow angle regions in the circumferential direction are exactly in accordance with the initial swirl direction of the burners. The persistence of initial vortex structures until outlet emphasizes the significance of a proper configuration of the SHC swirlers regarding their rotational direction.

The cold flow analysis showed that the circumferentially adjacent burners must have corotating swirls to avoid large deformation of the recirculation bubbles and nonuniform flow at the SHC outlet. Therefore, only the DA4 and DA6 types are selected for further analyses via reacting flow simulations. Both of them feature circumferentially corotating swirls where, in the DA4 type, the radially adjacent swirls are corotating as well, whereas in the DA6 the radially adjacent swirls are counterrotating.

In the following reacting flow simulations, the effects of gas expansion and acceleration due to the combustion on the SHC flow are explored.

Reacting Flow

The combustion of a lean-premixed mixture of methane and air at an equivalence ratio of $\phi = 0.5$ is taken as the basis for all simulations. For similarity reasons, the mass flow of the fuel–air mixture is the same as the mass flow of air of the nonreacting flow simulations, and the cooling air is not considered.

The predicted fields of the velocity $U_{x,\beta}$ and the temperature for the DA4 and DA6 types are shown in Fig. 6.12. On the C–C planes, the position of the swirling jets and the associated patterns of the high-velocity zones are similar to the nonreacting flow field. The main difference is, however, that due to combustion the swirling jets exhibit higher $U_{x,\beta}$ values, and the swirling cores are compacter.

The symmetric placement of the swirling jets and the flame fronts of the DA6 type lead to a high $U_{x,\beta}$ zone at the middle height of the combustor. Because of the radial balance of the flow structure in the whole combustor, in contrast to the DA4 type, the shape and position of the highly angled jet are maintained downstream until the combustor exit. The latter is marked with the large arrow in Fig. 6.12.

The exit flow patterns of DA4 and DA6 types are presented in Fig. 6.13, where the analogy to the velocity fields on the C–C planes can be recognized. Here, the relation of the flow angle and

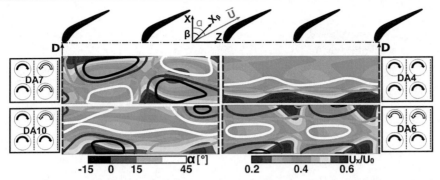

Figure 6.11: Nonreacting flow. Contour plots of the time-averaged velocity field with isolines of the time-averaged flow angle on D–D planes.

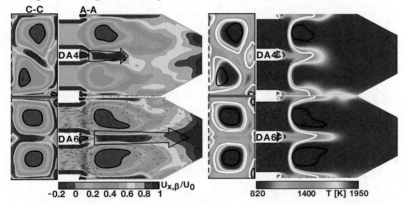

Figure 6.12: Reacting flow. Contour plots of the time-averaged velocity and temperature fields. Black isolines correspond to $U_{x,\beta} = 0$. White isolines correspond to progress variable $\theta = 0.5$, representing the flame front. Arrows highlight the highly angled flow at the mid-height of the flame tube.

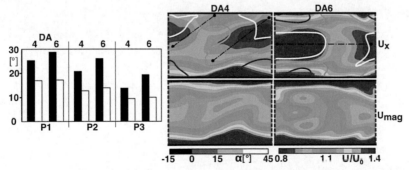

Figure 6.13: Reacting flow. Bar diagram of flow angle and its standard deviation in the DA4 and DA6 configurations (left). The evaluation planes P1–P3 are defined in Fig. 6.9. Contour plots of the velocity field with isolines of flow angle at D–D planes. Dash-dotted lines highlight the orientation of the zones of large flow angles.

velocity fields should be discussed, again to highlight the fact that the SHC outlet flow condition is primarily determined by the initial swirl configuration.

Accordingly, it can be seen that the zones of a large flow angle α, as marked with dash-dotted lines in Fig. 6.13, emerge at the similar spatial positions as the far upstream placed zones of high $U_{x,\beta}$. The latter were arranged dependent on the interaction of the adjacent swirling jets due to their rotational direction.

Furthermore, the large α zones exhibit low axial velocities U_x, which can be explained by the definition of the flow angle

$$tan(\alpha) = \frac{U_z}{U_x} \qquad (6.10)$$

Accordingly, as schematically shown in Fig. 6.13, an increased flow angle, by a constant magnitude of the velocity vector, is associated with lower axial and higher circumferential velocities. The magnitude of the velocity vector, as shown in Fig. 6.13, exhibits a high circumferential uniformity at the outlet. The higher velocity magnitudes next to the liner walls refer to the upstream aerodynamic blocking effect of the recirculation zones of the swirling flames at the primary combustion zone.

To conclude, the large flow angle region at the middle height of the DA6 corresponds to the similarly formed high $U_{x,\beta}$ zone at the same position at the inlet. The latter is the direct result of the interaction of radially adjacent counterrotating swirling jets.

In contrast to the DA6, the placement of the swirling jets and so the high $U_{x,\beta}$ regions of the DA4 type are radially asymmetric (see the C–C plane in Fig. 6.12). This leads to an imbalanced flow. The high $U_{x,\beta}$ region at the mid-height of the DA4 type is decayed due to the balancing flow motion and does not reach the combustor exit. However, analogous to the DA6 type, at the exit plane, the influence of the initial high $U_{x,\beta}$ zone on the flow pattern is still visible due to the inclined large flow angle zones as highlighted in Fig. 6.13 by dash-dotted lines.

The bar diagram in Fig. 6.13 shows the evolution of the average flow angle of the combustor exit nozzle, on the P1–P3 planes as defined in Fig. 6.9. In contrast to the nonreacting flow, the exit mean flow angle of the DA6 is higher than the DA4 type. The high $U_{x,\beta}$ zone at the middle height of the DA6 type is augmented due to the combustive gas expansion and dominates the whole combustor flow. It establishes a radially symmetric flow pattern with large positive flow angles at the exit of the L0-14DA45 SHC type.

Summary of the SHC Flow Kinematics

The swirl rotational direction of the circumferentially adjacent burners is a principal parameter of the SHC design. It determines the entire combustor flow. In conventional annular combustors, at each cross section normal to the burner axes, the tangential and radial velocities of the adjacent swirling flames are balanced both in co- and counterrotating constellations. Hence, the rotational direction of adjacent flames is rather a tuning factor regarding the mixing and combustion performance. In the SHC, however, because of circumferential asymmetry brought about by the sidewalls, there is no balancing effect by the circumferentially adjacent flames.

Accordingly, counterrotating swirls of circumferentially adjacent burners must be avoided due to the destructive interaction of the flow at the segment boundaries. Furthermore, it is shown that the inlet vortex structures persist until the combustor exit and determine the exit flow patterns. For example, the DA6 type exhibits large flow angles at the middle height and low flow angles near the liner walls, which agree with the initial swirl directions. Likewise, the alternating regions of positive and negative flow angles at the exit of the DA7 type can be attributed to its segment-wise counterrotating swirls at the inlet.

As a result of the nonreacting flow analysis, the DA4 and DA6 types have been selected for reacting flow simulations. The DA4 type is selected because of the uniform circumferential flow pattern and the maximum flow angle at the exit. The DA6 type is selected because of the radially symmetric flow pattern in the primary zone and at the combustor exit.

The reacting flow simulations showed that the counterrotating swirls of radially adjacent burners are a further essential feature for a DA–SHC. The corresponding DA6 type exhibits at the exit plane an almost symmetric flow pattern with the largest mean flow angle of $\bar{\alpha} \approx 20°$. Furthermore, it features large flow angles at the middle height and low at the endwalls, which is expected to be advantageous regarding the aerodynamics and heat transfer of the NGV.

In the DA6 type as the superior SHC configuration, however, the exit mean flow angle is lower than half of its initial value at the combustor inlet. The reasons and origins of this unwanted diminution of the initial high angular momentum flux cannot be explained by the kinematic flow analysis conducted in this section. Of particular interest is to track the contribution of the pressure and viscous forces on this inertial flow phenomenon, which is the subject of the following section.

6.2.3   The SHC Flow Dynamics

By the former analysis of the flow kinematics, it was revealed that the configuration of the rotational direction of the adjacent swirling flows is a determining factor for the SHC flow. However, the superior DA4 and DA6 swirl configurations still exhibit very low mean flow angles and nonuniform flow patterns at the flame tube exit planes. The underlying physics of those flow conditions are elucidated by the analysis of the flow dynamics based on the integral balance of angular momentum as introduced in Sec. 4.1.

The development of the different terms of Eq. (4.2) for nonreacting and reacting flow through the L0-14DA45 SHC types is shown in Fig. 6.14. The main phenomenon observed is the strong decrease of angular momentum flow L_{out} in the region of staggered walls. Almost reciprocally, the pressure torque L_p induced by the wall pressure forces increases in that region. The friction torque L_τ increases slightly in the vicinity of the burners, which is due to the wall shearing in this region. However, in both cold and combustive cases, the sum of all viscous effects is about one order of magnitude smaller than angular momenta due to inertial and pressure forces.

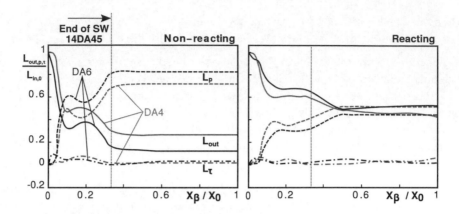

Figure 6.14: Integral balance of angular momentum in the SHC models as defined in Eq. (4.2). $L_{in,0}$ refers to initial angular momentum flow into an L0-7SA45-SA1.

In that respect, it is necessary to investigate the parameters that are connected to the contribution of the pressure torque to angular momentum balance. These are, on the one hand, the surface area of the sidewall, which multiplied by the pressure yields the force. And on the other hand, the magnitude and distribution of pressure on the sidewalls, which is mainly controlled by the swirl configuration of the adjacent burners. The effect of the sidewall's surface area was investigated in Ariatabar et al. (2016, 2017) and is discussed briefly below. The pressure magnitude and distribution effects are analyzed in detail and build the basis for flow control and improvement of the SHC in the next section.

The course of angular momentum flow shows for both the DA4 and DA6 types significant changes between the inlet and the end of the sidewall. This suggests the *size* of the sidewall being the main influence parameter controlling angular momentum flow in the SHC. The total area of the sidewalls in an SHC, when designed by the *L*0 scaling law based on Eq. (6.6), is exclusively determined by the tilting angle β

$$A_{sw,\beta} = N_{seg} H_{\beta} P_{sw} = N_{seg} \frac{H_{rcc}}{cos\beta} \frac{\pi D_{rcc}}{N_{seg}} sin\beta = A_{dm,rcc} tan\beta \qquad (6.11)$$

As an example, the total area of sidewalls in L0-14DA45 is twice as large as that of the L0-14SA25 with $A_{sw,45}/A_{sw,25} \approx 2$. The L0-14SA25 case was investigated by Ariatabar et al. (2016). It was found that the decrease of angular momentum flow is considerably lower than that of the L0-14DA45. The reason was the lower pressure reaction torque exerted by the smaller sidewalls. On the other hand, the smaller tilting angle of the burners is associated with lower initial angular momentum flow L_{in}. Therefore, even though the L0-14SA25 type delivered a *relatively* high outlet angular momentum flux, the major benefits of the SHC will be missed by this type, namely the axial shortening of the combustor and the reduction of the NGV due to a highly angled flow field in the flame tube. Hence, the focus must be laid on remedying the shortcomings that arise from the large tilting angles of the burners in the SHC such as those of the L0-14DA45 type.

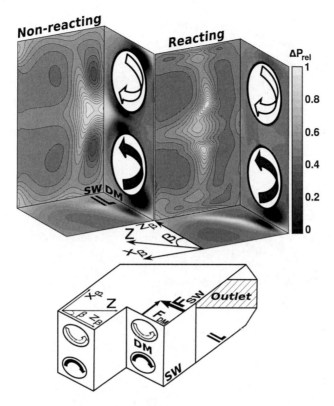

Figure 6.15: Contour plots of relative pressure field at the dome (DM), sidewall (SW), and inner liner (IL) wall in a DA6 configuration (top). Schematic illustration of asymmetric pressure reaction forces exerted by the sidewall F_{SW} and the dome F_{DM} and the corresponding deflection of the flow (bottom).

In the DA6 type, the pressure torque of nonreacting flow is higher than in the DA4, as plotted in Fig. 6.14. This highlights once more the sensitivity of the entire helical flow structure of the SHC to the rotational direction of the swirl of the adjacent burners. Accordingly, the reaction pressure torque induced by the nonreacting swirling flows of the DA6 type is larger than that of the DA4 type. This can be deduced from Fig. 6.7. The swirl configuration of the DA6 type establishes an augmented tangential flow at the mid-height directed toward the sidewall. This flow stagnates and builds a high-pressure region at the sidewall, which in turn exerts a high reaction torque on the swirling flows. In contrast, in the DA4 type as schematically indicated by the arrows in Fig. 6.7, the tangential velocity components of the corotating swirls compensate each other at the mid-height. Thus, the stagnation on the sidewalls is less intensive and the accompanying pressure torque is weaker than that of the DA6 type.

Moreover, it is apparent that the effects of individual swirl configurations are substantially dependent on whether the combustion takes place or not. Both configurations exhibit in the reacting cases a higher angular momentum flux at the outlet. Because of the thermal expansion due to the heat release, the flow accelerates parallel to the sidewalls. Thus, the stagnation of the swirling jets on the sidewalls becomes less intensive in the reacting case.

The effect of combustion is, however, more apparent for the DA6 than for the DA4 type. The reason is the compacter swirl cores in the reacting case (compare Fig. 6.7 and 6.12). Thus, the resulting tangential flow at the mid-height is weaker than the nonreacting case. This leads to a smoother impingement of the DA6 swirling jets on the sidewalls. Consequently, the stagnation sidewall pressure is significantly weaker than the nonreacting case. Moreover, the reaction pressure torque is smaller as plotted in Fig. 6.14.

The associated pressure distribution on the walls of the DA6 type is presented in Fig. 6.15. It can be seen that in the reacting case the relative pressure at the sidewall exhibits a lower level than in the nonreacting case. Moreover, the pressure distribution between the sidewall and dome is more uniform in the reacting case. As a consequence, the resulting pressure force on the flow is smaller. Thus, the swirling cores are less deflected in the axial direction, and the outflowing angular momentum is higher than in the nonreacting case as shown in Fig. 6.14.

Summary of the SHC Flow Dynamics

The analysis of the fluid dynamics of the SHC elucidated that the pressure distribution on the sidewalls governs the flow in the primary combustion zone. Accordingly, the friction and turbulent mixing phenomena are found to have marginal effects on the flow structure. It is shown that the swirling jets induce an asymmetric pressure distribution on the sidewall and the combustor dome. The resulting reaction wall pressure forces give rise to a torque in the counterrotating direction to the turbine. As a consequence, the flow is deflected toward the axial direction.

The torque exerted by the sidewall pressure on the flow is influenced by

1. The size of the sidewalls

2. The swirl configuration

3. Combustion

Moreover, it was found that the size of the sidewalls is a function of the tilting angle of the burners as given by Eq. (6.11). The tilting angle, on the other hand, must satisfy the similarity rules and also be maximized to achieve a significant shortening of the combustor length and reduction of the number of the NGV. According to the parametric study presented in Fig. 6.2, the most appropriate configuration for the SHC is a tilting angle of $\beta = 45°$. Consequently, it must come to terms with the associated large sidewalls.

Furthermore, for this tilting angle, the analysis of the kinematics of the flow in Sec. 6.2.2 showed that the DA6 type is the superior configuration. The subsequent dynamic flow analysis showed that the corresponding DA arrangement of the burners with radially counterrotating swirls induce a radially symmetric pressure field on the sidewalls. Because of combustion, the pressure level on the sidewalls is reduced, and the associated reaction torque is decreased. Hence, within the whole flame tube, a radially symmetric flow structure is established. Furthermore, the DA6 type exhibited the highest mean exit flow angle of all investigated SHC configurations. This, however, is only about $\bar{\alpha} \approx 20°$ with a low circumferential uniformity.

The tremendous diminution of the SHC angular momentum inflow might be obviated based on the insight gained from the global flow analysis. As a consequence, a high angular momentum outflow, as well as an improvement of the aerothermal uniformity of the flow field at the flame tube exit plane, is expected.

In the next section, the aforementioned hypothesis will be tested. The target will be the reduction of the reaction pressure torque induced by the sidewall. It will be pursued by means of flow field manipulation via contouring the confining walls. For that, a generic contoured sidewall for a 14DA45-DA6 is proposed, which aims at the radial deflection of the impinging swirling jets and prevention of high stagnation pressure regions. The corresponding fluid dynamics of the predicted flow fields are assessed based on the integral balance equation for angular momentum as introduced in Sec. 4.1.

6.3 Controlling the SHC Flow via Wall Contouring

In this section, it will be investigated if generic contouring of the sidewall can prevent the significant decline of angular momentum flow in the primary combustion zone of the SHC. It will be further examined, if with this approach the aerothermal uniformity of the combustor exit flow can be improved.

The primary goal of the sidewall contouring is the reduction of the pressure torque exerted on the flow. As shown in Fig. 6.15, because of the specific swirl configuration of the DA6 type, a high pressure zone is established at the mid-height of the sidewall. This was identified as the main parameter diminishing the initial angular momentum flow. The latter may be averted by avoiding the steep impingement of the swirling jets on the sidewall. This can be carried out by guiding the swirling jets over a contoured wall geometry. The idea is to establish a low wall pressure field as homogeneous as possible with smooth gradients. It is pursued via local acceleration or deceleration of the flow by decreasing or increasing the cross-sectional area of the flame tube through contouring of the liner walls.

The secondary goal is to reduce the circumferential component of the wall pressure force regardless of its magnitude. At a first glance, this can be achieved via a pure shape optimization of the wall contour geometry no matter how the flow field is developed around. However, because the fluid dynamics of the swirling flows and so the resulting pressure fields are coupled with the shape of the sidewall, an optimization with an exclusive geometrical target function may not be expedient. Hence, considering the complexity of the asymmetrically bounded swirling flows in the SHC, the optimization would be a demanding iterative process and beyond the scope of the present thesis.

Nevertheless, to study the effect of such a wall contouring on the fluid dynamics of the combustor flow, a starting point would be a generic contouring considering both objectives mentioned above separately.

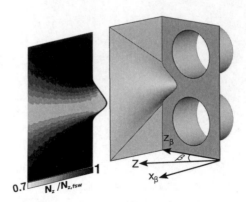

Figure 6.16: Contoured sidewall of the SHC type 14DA45 and a schematic illustration of the circumferential components of the surface normals.

To that end, to accelerate the flow and reduce the wall pressure to pursue the primary objective of the liner wall contouring, the proposed sidewall should exhibit an elevation at the mid-height, in the form of a ridgeline parallel to the burner axes as shown in Fig. 6.16.

Furthermore, the secondary objective is followed at first by minimization of the area-weighted average value of the circumferential component of the wall surface normal vectors, $\bar{N}_z$.

The reduction of the wall pressure field in the design shown in Fig. 6.16 is pursued based on the engineering judgment of the flow kinematics as well as the reduction of the circumferential component of the surface normal vectors based on the following pure geometrical considerations.

Hereafter, the 14DA45-DA6 with a flat sidewall is denoted by FSW and its modification with a contoured sidewall is denoted by CSW.

Fig. 6.16 shows a contour plot of the normalized N_z. The reference value is the constant $N_{z,fsw}$ of the FSW. Accordingly, on the ridge of the CSW the N_z values are smaller than those of the FSW. The reason is that the surface normal vectors on the ridge are decomposed into the axial, circumferential, and additionally also into a radial component. Thus, in a proper design, the circumferential component can be minimized. The normalized circumferential component of the surface normal vector is given by

$$N_{z,norm} = \frac{N_z}{N_{z,fsw}} = \frac{N_{z,\beta}\,cos(\beta) + N_{x,\beta}\,sin(\beta)}{N_{z,\beta,fsw}\,cos(\beta)} \tag{6.12}$$

For the current generic CSW, an area-weighted average of $\bar{N}_{z,norm} = 0.92$ is achieved. This value shows that in the case of a *uniform pressure field*, a reduction of 8% of the circumferential pressure force can be achieved, just by a geometrical modification of the sidewall.

The proposed generic contoured wall should only demonstrate the potential of wall contouring to manipulate the flow structure particularly of the SHC and generally in gas turbine combustors. The effectiveness of this approach is investigated via assessment of the flow kinematics and dynamics in the following sections.

6.3.1 Effect of the Wall Contouring on the Flow Kinematics

In Fig. 6.17, the contours of the velocity $U_{x,\beta}$ are plotted for the A–A planes of the CSW and FSW types. The contoured ridge on the sidewall forces the swirling jets into the radial direction away from each other. This can be recognized by the larger angle between the swirl axes linked to the isohypses of the contoured sidewall. The spatial separation of individual recirculation bubbles by the ridge reduces the interaction of the radially adjacent swirling jets. Consequently, the CSW type exhibits larger regions of high $U_{x,\beta}$ values both at the mid-height and next to the liner walls. These regions are indicators of the highly angled flow and are responsible for a large mean flow angle at the outlet of the CSW type.

In the CSW type, the mean flow angle is almost doubled compared with the FSW design. Thereby, on the one hand, a larger area of highly angled flow is established at the mid-height of the NGV. On the other hand, the negative flow angle zones near the inner and outer liners are suppressed. This can be explained by the schematic depiction of the swirling jets in Fig. 6.17. The axes of swirling jets of the CSW type are tilted toward the inner and outer liner walls. Accordingly, the velocity components in the negative circumferential direction are exposed to a higher shear stress rate, and thus, are decaying more intensively.

The reacting flow fields are illustrated in Fig. 6.18 by velocity contour plots. The characteristic deflection of the swirl flames of the FSW type can be recognized by the deformation of the recirculation bubbles in the radial (A–A planes) and circumferential (B–B planes) directions. In the previous discussions, the flow field was described based on the asymmetric pressure distribution on the sidewall and the dome. Based on the same physics, the deflection and deformation of IRZ can also be explained by the Coandă effect (Coanda, 1936, Reba, 1966, Wille and Fernholz, 1965).

As in the nonreacting case, the CSW type provides not only a higher mean flow angle at the exit but also a more uniform flow in the circumferential direction. This is particularly advantageous for the heat transfer at the downstream NGV. The vortex pairs, as schematically illustrated in the temperature plots, at their cores isolate the temperature hot spots from the surrounding fluid and prevent effective mixing. More details about the underlying physics of temperature and scalar mixing in vortex cores of isothermal and reacting flows can be found in Alain and Candel (1988), Basu et al. (2007), Laverdant and Candel (1989). In addition to transporting hot spots, the vortex structures locally distort the endwall heat transfer at NGV and reduce the film cooling effectiveness as shown recently by Werschnik et al. (2017a,b), among others.

As discussed in Sec. 6.2.2, the interaction of radially adjacent swirling flames can establish particular high-velocity jets in the flame tube. The shape and position of the jets are foremost dependent on the rotational direction of the adjacent swirling flames. Around these jets, regions of low-pressure emerge. In consequence of the imbalance in the low-pressure zones, a pressure force will be exerted on the fluid from the weaker toward the stronger jets. Hence, the recirculation zones will be deformed in accordance with the shape and position of the jets.

Contouring of the liner walls is an effective measure to counteract the imbalance of the high-velocity regions in the primary combustion zone of the SHC. As shown in Fig. 6.20, a generic

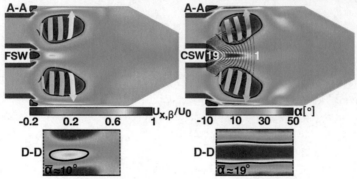

Figure 6.17: Nonreacting flow. Contour plots of the time-averaged velocity field with isolines of $U_{x,\beta} = 0$ on the A–A planes. The isohypses on the CSW type indicate the wall contouring from 1mm to 19mm. The radial deflections of the swirling jets are highlighted schematically with spiral arrows (top). Contour plots of time-averaged flow angle field with isolines of $\alpha = 30°$ on the D–D planes (bottom). Mean values $\bar{\alpha}$, correspond to the physically consistent average flow angle. Dashed lines indicate the periodic boundaries.

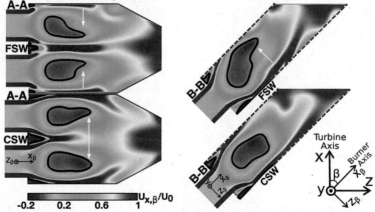

Figure 6.18: Reacting flow. Contour plots of the time-averaged velocity field with isolines of $U_{x,\beta} = 0$ on the A–A (left) and the B–B planes (right). White arrows represent the Coandă pressure forces in consequence of an imbalance of high-velocity regions. Dashed lines indicate the periodic boundaries.

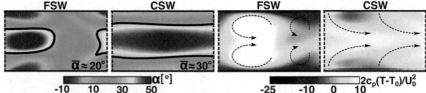

Figure 6.19: Reacting flow. Contour plots of the time-averaged flow angle field with isolines of $\alpha = 30°$ on D–D planes (left). Mean values $\bar{\alpha}$, correspond to the physically consistent average flow angle. Contour plots of time-averaged temperature field on D–D planes with dashed arrows indicating the vortex systems (right). Dashed lines indicate the periodic boundaries.

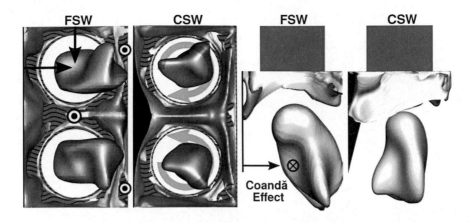

Figure 6.20: Reacting flow. Front (left) and top (right) view of isosurfaces of the time-averaged velocity $U_{x,\beta} = 0$. Arrows and $\otimes$ sign in FSW represent the Coandă pressure forces in consequence of high-velocity jets located at zones marked by $\odot$ signs. Domes are hatched by wavy patterns. Transparent arrows show the rotational directions of the swirling jets.

contouring of the sidewall will bring about a more balanced and symmetric flow field. Accordingly, the local contraction of the flow cross section will partially prevents the large *corner* recirculation zones associated with the sudden expansion of the swirling jets. Instead, the jets are smoothly guided over the ridge on the sidewall, and the corner dead-water zones are distributed more uniformly over the dome and sidewall area. This, in turn, is associated with a more symmetric flow structure and balanced pressure distribution. Thus, the swirl flames are less deformed and not deflected away from the burner axes.

6.3.2 Effect of the Wall Contouring on the Flow Dynamics

In this section, analogously to Sec. 6.2.3, the deflection of the swirl axes is analyzed by evaluating the driving forces. Based on the integral balance of angular momentum, as presented in Sec. 4.1, the evolution of angular momenta in the FSW and CSW types are plotted in Fig. 6.21.

For the nonreacting cases, the wall pressure forces exerted by the sidewall on the swirling flow in both the FSW and CSW types are responsible for the notable steep decline of angular momentum flow at the inlet of the flame tube. The pressure torque L_p increases in that region, whereas the flux of angular momentum L_{out} decreases almost reciprocally. Downstream of the sidewalls, there are no surfaces where external pressure forces can be imposed onto the flow. Downstream of the sidewalls, the Coandă pressure forces due to the imbalance of high $U_{x,\beta}$ regions are responsible for the further increase of the torque induced by the wall pressure.

As shown on the B–B planes in Fig. 6.18, the imbalance of the jets in FSW is augmented in the reacting case, whereas the CSW maintains a symmetric and balanced flow structure. The contributions of the friction torque L_τ are in all cases negligible and about one order of magnitude

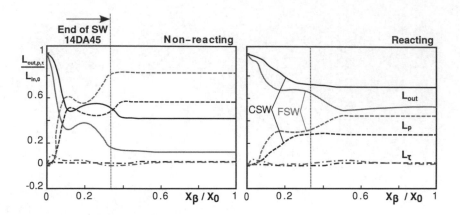

Figure 6.21: Integral balance of angular momentum in the SHC models as defined in Eq. (4.2). $L_{in,0}$ refers to initial angular momentum flow into an L0-7SA45-SA1.

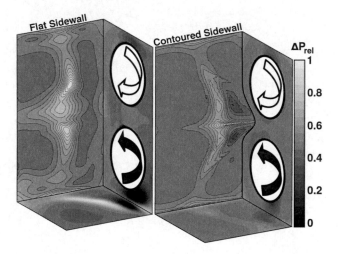

Figure 6.22: Reacting flow. Contour plots of the time-averaged relative pressure field in an SHC with a Flat Sidewall and a Contoured Sidewall.

smaller than the inertial and pressure torques.

The pressure at the walls of the primary combustion zones of the FSW and CSW types are plotted in Fig. 6.22. In contrast to the FSW type, the swirling jets at the inlet of the flame tube in the CSW type are smoothly guided over the contoured ridge on the sidewall. In consequence, the pressure distribution is more homogeneous and extreme low- and high-pressure zones disappeared. Hence, the resulting pressure force in the adverse circumferential direction is significantly reduced.

Summary of the Flow Control via Wall Contouring in the SHC

It is shown that liner wall contouring is a robust and effective approach to control the SHC flow with high potential. The SHC angular momentum outflow could be increased by almost 50% via a simple generic contouring of the sidewalls.

Moreover, it is shown that the local manipulation of the flow in the primary combustion zone directly affects the outlet flow structure. Accordingly, the circumferential uniformity of the velocity and temperature fields at the SHC exit plane could be significantly increased via manipulation of the wall pressure field in the primary combustion zone.

The principal approach of controlling the combustor flow by manipulating the wall pressure field can be applied in many conventional combustion devices. The demonstrated contouring of the liner wall can open a new horizon in combustor design engineering. The aerothermal nonuniformities at the flame tube exit plane can be tackled at their early formation stages in the primary combustion zones.

Accordingly, an effective tool, in the future, to enhance the combustor–turbine interaction, as well as the NGV endwall cooling, might be the aerothermal tailoring of the combustor exit flow pattern through contouring of the liner walls.

7 Summary and Conclusions

The main concern of this thesis is the investigation of an annular gas turbine combustor with angular air supply from the aerothermal point of view, based on analytical–numerical methods.

In the introductory chapter, the study is motivated by the increasing demand for high-efficiency and low-emission gas turbines. One approach to pursue this objective is the enhancement of the coupling and interaction of the combustor with the compressor and the turbine. In this context, an innovative solution is offered by the combustor concept that is the subject of the present study.

The principal idea of the so-called *Short Helical Combustor* is the utilization of angular momentum of the compressor airflow discharged from the last rotor cascade. Hence, the number of nozzle guide vanes and the related cooling air demand can be reduced. Furthermore, a helical flow pattern would be established in the flame tube. With a "constant residence time" scaling of the SHC, this helical flow structure can be adopted to shorten the axial length of the flame tube and so the entire engine. All these features would lead to the increased overall efficiency of the gas turbine, and the decrease in its specific fuel consumption and pollutants' emissions.

Starting with Seippel (1943), there have been numerous further patents filed by almost all major gas turbine manufacturers describing various implementations of this combustor concept. However, the SHC has never been realized in civil gas turbines. The main reason was probably that the manufacturers recognized a complex and distorted flow field in the combustor prototypes. Moreover, there is no fundamental research available that addresses the essential flow features and optimization possibilities of such an unconventional design with asymmetric inflow and geometric boundaries.

With this being the primary objective of the present study, in Chapters two to six, the research methodology is defined, and the necessary tools in a numerical framework are developed and validated.

In the last chapter, the SHC is investigated via analytical and numerical methods. It is shown that a double annular configuration with a tilting angle of $\beta = 45°$ can satisfy the most relevant similarity and scaling rules. However, at the combustor exit, an average flow angle of only $\bar{\alpha} \approx 20°$ could be achieved. Besides, the velocity and temperature fields at the outlet exhibited a low circumferential uniformity.

The integral balance of angular momentum is employed to analyze the SHC flow. It is shown that the tremendous decrease of the flow angle is primarily caused by a high reaction wall pressure torque due to the asymmetric distribution of the stagnation pressure at the sidewalls and the combustor dome. In this regard, a generic design improvement is investigated. The principal design, patented by (Ariatabar et al., 2019), features a ridge on the sidewall aiming at guidance of the swirling jets and preventing the extreme wall pressure zones. Consequently, by contouring the flame tube walls in the primary combustion zone:

1. The shapes and positions of the swirling flames could be modified purposefully.

2. The average outlet flow angle was increased by 50% compared with the reference case without contoured flame tube walls.

3. With the SHC featuring contoured flame tube walls, the turning angle and the number of required NGV are reduced by $\approx 40\%$

4. The proposed double annular SHC configuration ensures an axial shortening of 30% compared with an equivalent conventional combustor. This could be achieved exclusively due to the staggered arrangement of the flames together with appropriate scaling of the burners and the flame tube.

The improvements mentioned above are associated explicitly with the SHC. However, the contouring of combustor walls has a great potential to improve the major characteristics of conventional swirl-stabilized gas turbine combustors too. Accordingly, the static and dynamic combustion stabilities might be enhanced by manipulating the shape, size, and location of the swirl flames.

Furthermore, the combustor–turbine interaction might be improved regarding the uniformity of the velocity and temperature fields at the outlet plane. Notably, the endwall heat transfer and film cooling effectiveness at NGV might be enhanced.

The generic examples presented in this dissertation should only shed light on the potential of the flow control in gas turbine combustors by contouring of the flame tube walls. With respect to the new possibilities offered by additive manufacturing, wall contouring might shape the future of the gas turbine combustors.

References

Abdel-Gayed, RG, Bradley, D, Hamid, MN and Lawes, M (1985): *Lewis Number Effects on Turbulent Burning Velocity*. In: *Symposium (International) on Combustion*, vol. 20, pp. 505–512. Elsevier.

Abdel-Gayed, RG, Bradley, D and Lung, FKK (1989): *Combustion Regimes and the Straining of Turbulent Premixed Flames*. Combustion and Flame, vol. 76, pp. 213–218.

Alain, M and Candel, SM (1988): *A Numerical Analysis of a Diffusion Flame-Vortex Interaction*. Combustion Science and Technology, vol. 60, pp. 79–96.

Anderson, J (1995): *Computational Fluid Dynamics*. Computational Fluid Dynamics: The Basics with Applications. McGraw-Hill Education, New York.

Andrews, GE and Bradley, D (1972): *The Burning Velocity of Methane-Air Mixtures*. Combustion and Flame, vol. 19, pp. 275–288.

Angelberger, C, Veynante, D and Egolfopoulos, F (2000): *LES of Chemical and Acoustic Forcing of a Premixed Dump Combustor*. Flow, Turbulence and Combustion, vol. 65, pp. 205–222.

Ariatabar, B, Koch, R and Bauer, HJ (2014): *SHC Brennkammer: Entwicklung Einer Ringbrennkammer mit Spiralförmiger Flammenanordnung*. In: *Informationstagung Turbomaschinen*, vol. R567-2014, pp. 63–92. Forschungsvereinigung Verbrennungskraftmaschinen e.V.

Ariatabar, B, Koch, R and Bauer, HJ (2015a): *SHC Brennkammer: Entwicklung Einer Ringbrennkammer mit Spiralförmiger Flammenanordnung*. Abschlussbericht 1067-2015, Forschungsvereinigung Verbrennungskraftmaschinen e.V.

Ariatabar, B, Koch, R, Bauer, HJ and Negulescu, DA (2015b): *Short Helical Combustor: Concept Study of an Innovative Gas Turbine Combustor With Angular Air Supply*. Journal of Engineering for Gas Turbines and Power, vol. 138, pp. 31503,1–31503,10.

Ariatabar, B, Koch, R and Bauer, HJ (2016): *Short Helical Combustor: Dynamic Flow Analysis in a Combustion System With Angular Air Supply*. Journal of Engineering for Gas Turbines and Power, vol. 139, pp. 41505,1–41505,8.

Ariatabar, B, Koch, R and Bauer, HJ (2017): *Short Helical Combustor: Flow Control in a Combustion System With Angular Air Supply*. Journal of Engineering for Gas Turbines and Power, vol. 140, pp. 31507,1–31507,6.

Ariatabar, B, Koch, R and Bauer, HJ (2019): *Gasturbinenbrennkammeranordnung*. Deutsches Patent- und Markenamt. Patent Nr.: DE 10 2017 100 984 B4.

Aris, R (2012): *Vectors, Tensors and the Basic Equations of Fluid Mechanics*. Courier Corporation, Englewood Cliffs - New Jersey.

Arrhenius, S (1889): *Über die Reaktionsgeschwindigkeit bei der Inversion von Rohrzucker Durch Säuren*. Zeitschrift für physikalische Chemie, vol. 4, pp. 226–248.

Bachelor, GK (1967): *An Introduction to Fluid Mechanics.*   Cambridge University Press, Cambridge.

Bardina, JE, Huang, PG and Coakley, T (1997): *Turbulence Modeling Validation,Testing, and Development*. Tech. rep., NASA Ames Research Center. Technical Memorandum 110446.

Bärow, E, Koch, R and Bauer, HJ (2013): *Comparison of Oscillation Modes of Spray and Gaseous Flames.*

Basu, S, Barber, TH and Cetegen, BM (2007): *Computational Study of Scalar Mixing in the Field of a Gaseous Laminar Line Vortex*. Physics of Fluids, vol. 19, pp. 53601,1–53601,13.

BDL (2017): *Klimaschutz Report 2017*. techreport, Bundesverband der Deutschen Luftverkehrswirtschaft.

Beér, JM and Chigier, NA (1972): *Combustion Aerodynamics*. Fuel and energy science series. Krieger Publishing Company, Malabar - Florida.

Benjamin, TB (1962): *Theory of the Vortex Breakdown Phenomenon*. Journal of Fluid Mechanics, vol. 14, pp. 593–629.

Benjamin, TB (1967): *Some Developments in the Theory of Vortex Breakdown*. Journal of Fluid Mechanics, vol. 28, pp. 65–84.

Bilger, RW, Pope, SB, Bray, KNC and Driscoll, JF (2005): *Paradigms in Turbulent Combustion Research*. Proceedings of the Combustion Institute, vol. 30, pp. 21–42.

Billant, P, Chomaz, JM and Huerre, P (1998): *Experimental Study of Vortex Breakdown in Swirling Jets*. Journal of Fluid Mechanics, vol. 376, pp. 183–219.

Bird, RB, Stewart, WE and Lightfoot, EN (2007): *Transport Phenomena*. A Wiley International edition. John Wiley & Sons, Hoboken - New Jersey, 2nd edn.

Bockhorn, H, Mewes, D, Peukert, W and Warnecke, HJ (2009): *Micro and Macro Mixing: Analysis, Simulation and Numerical Calculation*. Heat and Mass Transfer. Springer, Berlin.

Bohan, BT and Polanka, MD (2013): *Analysis of Flow Migration in an Ultra-Compact Combustor*. Journal of Engineering for Gas Turbines and Power, vol. 135, pp. 51502.

Borghi, R (1985): *On the Structure and Morphology of Turbulent Premixed Flames*. In: *Recent Advances in the Aerospace Sciences*, pp. 117–138. Springer.

Boussinesq, J (1877): *Théorie De L' Écoulement Tourbillant*.   Mem. Présentés par Divers Savants Acad. Sci. Inst. Fr, vol. 23, pp. 6–5.

Bradley, D (1992): *How Fast Can We Burn?*. In: *Symposium (International) on Combustion*, vol. 24, pp. 247–262. Elsevier.

Bradley, D, Gaskell, PH, Gu, XJ, Lawes, M and Scott, MJ (1998): *Premixed Turbulent Flame Instability and NO Formation in a Lean-Burn Swirl Burner*. Combustion and Flame, vol. 115, pp. 515 – 538.

Bray, KNC (1980): *Turbulent Flows with Premixed Reactants*. In: *Turbulent Reacting Flows*, pp. 115–183. Springer.

Bray, KNC and Moss, JB (1977): *A Unified Statistical Model of the Premixed Turbulent Flame*. Acta Astronautica, vol. 4, pp. 291–319.

Burd, SW and Cheung, AK (2007): *Angled Flow Annular Combustor For Turbine Engine*. WO Patent App. PCT/US2006/007,898.

Buret, MR, Cazalens, MP and Hernandez, DH (2009): *Turbomachine With Angular Air Delivery*. US Patent 7,549,294.

Burmberger, S and Sattelmayer, T (2011): *Optimization of the Aerodynamic Flame Stabilization for Fuel Flexible Gas Turbine Premix Burners*. Journal of Engineering for Gas Turbines and Power, vol. 133, pp. 101501.

Burrus, DL, Johnson, AW, Roquemore, WM and Shouse, DT (2001): *Performance Assessment of a Prototype Trapped Vortex Combustor Concept for Gas Turbine Application*. In: *ASME Turbo Expo 2001: Power for Land, Sea, and Air*. American Society of Mechanical Engineers.

Cai, J, Jeng, SM and Tacina, R (2002): *Multi-Swirler Aerodynamics: Comparison of Different Configurations*. In: *ASME Turbo Expo 2002: Power for Land, Sea, and Air*, pp. 739–748. American Society of Mechanical Engineers.

Cary, AW, Darmofal, DL and Powell, KG (1998): *Evolution of Asymmetries in Vortex Breakdown*. In: *29th AIAA, Fluid Dynamics Conference*, pp. 2904,1–2904,9.

Chao, YC, Leu, JH, Hung, YF and Lin, CK (1991): *Downstream Boundary Effects on the Spectral Characteristics of a Swirling Flow Field*. Experiments in fluids, vol. 10, pp. 341–348.

Chase, MW, Curnutt, JL, Prophet, H, McDonald, RA and Syverud, AN (1975): *JANAF Thermochemical Tables, 1975 Supplement*. Journal of physical and chemical reference data, vol. 4, pp. 1–176.

Chen, JY (1997): *Stochastic Modeling of Partially Stirred Reactors*. Combustion Science and Technology, vol. 122, pp. 63–94.

Coanda, H (1936): *Device for Deflecting a Stream of Elastic Fluid Projected Into an Elastic Fluid*. US Patent 2,052,869.

Colin, O and Rudgyard, M (2000): *Development of High-Order Taylor-Galerkin Schemes for LES*. Journal of Computational Physics, vol. 162, pp. 338–371.

Colin, O, Ducros, F, Veynante, D and Poinsot, T (2000): *A Thickened Flame Model for Large Eddy Simulations of Turbulent Premixed Combustion*. Physics of Fluids, vol. 12, pp. 1843–1863.

Correa, SM (1998): *Power Generation and Aeropropulsion Gas Turbines: From Combustion Science to Combustion Technology*. Symposium (International) on Combustion, vol. 27, pp. 1793–1807.

Courant, R, Friedrichs, K and Lewy, H (1928): *Über die Partiellen Differenzengleichungen der Mathematischen Physik*. Mathematische annalen, vol. 100, pp. 32–74.

Cumpsty, NA and Horlock, JH (2006): *Averaging Nonuniform Flow for a Purpose*. Journal of Turbomachinery, vol. 128, pp. 120–129.

Damköhler, G (1940): *Der Einfluss der Turbulenz auf die Flammengeschwindigkeit in Gasgemischen*. Berichte der Bunsengesellschaft für physikalische Chemie, vol. 46, pp. 601–626.

Delery, JM (1994): *Aspects of Vortex Breakdown*. Progress in Aerospace Sciences, vol. 30, pp. 1–59.

Driscoll, JF, Chen, RH and Yoon, Y (1992): *Nitric Oxide Levels of Turbulent Jet Diffusion Flames: Effects of Residence Time and Damkohler Number*. Combustion and Flame, vol. 88, pp. 37–49.

Dunn-Rankin, D (2011): *Lean Combustion: Technology and Control*. Elsevier Science, London.

Dzung, LS (1971): *Konsistente Mittelwerte in der Theorie der Turbomaschinen für Kompressible Medien*. BBC-Mitt, vol. 58, pp. 485–492.

Echekki, T and Mastorakos, E (2010): *Turbulent Combustion Modeling: Advances, New Trends and Perspectives*, vol. 95. Springer, Berlin.

EPA (2017): *National Emission Standards for Hazardous Air Pollutants (NESHAP)*.

Escudier, M (1988): *Vortex Breakdown: Observations and Explanations*. Progress in Aerospace Sciences, vol. 25, pp. 189–229.

Escudier, MP and Keller, JJ (1985): *Recirculation in Swirling Flow - A Manifestation of Vortex Breakdown*. vol. 23, pp. 111–116.

Escudier, MP and Zehnder, N (1982): *Vortex-Flow Regimes*. Journal of Fluid Mechanics, vol. 115, pp. 105–121.

Escue, A and Cui, J (2010): *Comparison of Turbulence Models in Simulating Swirling Pipe Flowsturbulent channel flow up to*. Applied Mathematical Modelling, vol. 34, pp. 2840–2849.

Farokhi, S (2014): *Aircraft Propulsion*. John Wiley & Sons, Hoboken - New Jersey.

Farokhi, S, Taghavi, R and Rice, EJ (1989): *Effect of Initial Swirl Distribution on the Evolution of a Turbulent Jet*. AIAA J, vol. 27, pp. 700–706.

Favre, A (1965): *Equations des Gaz Turbulents Compressibles. Méthodes des Vitesses Moyennes, Méthodes des Vitesses Moyennes Pondérées par la Masse Volumique*. Journal de Mécanique, vol. 4, pp. 391.

Ferziger, JH and Peric, M (2012): *Computational Methods for Fluid Dynamics*. Springer, Berlin.

Fiorina, B, Vicquelin, R, Auzillon, P, Darabiha, N, Gicquel, O and Veynante, D (2010): *A Filtered Tabulated Chemistry Model for LES of Premixed Combustion*. Combustion and Flame, vol. 157, pp. 465–475.

Fox, RO (2003): *Computational Models for Turbulent Reacting Flows*. Cambridge Series in Chemical Engineering. Cambridge University Press, Cambridge.

Fritz, J, Kröner, M and Sattelmayer, T (2004): *Flashback in a Swirl Burner with Cylindrical Premixing Zone*. Journal of Engineering for Gas Turbines and Power, vol. 126, pp. 276–283.

Galpin, J, Naudin, A, Vervisch, L, Angelberger, C, Colin, O and Domingo, P (2008): *Large-Eddy Simulation of a Fuel-Lean Premixed Turbulent Swirl-Burner*. Combustion and Flame, vol. 155, pp. 247–266.

Gatski, TB and Speziale, CG (1993): *On Explicit Algebraic Stress Models for Complex Turbulent Flows*. Journal of fluid Mechanics, vol. 254, pp. 59–78.

Gepperth, S, Bärow, E, Koch, R and Bauer, HJ (2014): *Primary Atomization of Prefilming Airblast Nozzles: Experimental Studies Using Advanced Image Processing Techniques*. ILASS Europe, 26th Annual Conference on Liquid Atomization and Spray Systems, pp. 8–10.

Gicquel, LYM, Staffelbach, G, Cuenot, B and Poinsot, T (2008): *Large Eddy Simulations of Turbulent Reacting Flows in Real Burners: the Status and Challenges*. In: *Journal of Physics: Conference Series*, vol. 125, pp. 012029,1–012029,17. IOP Publishing.

Göttgens, J, Mauss, F and Peters, N (1992): *Analytic Approximations of Burning Velocities and Flame Thicknesses of Lean Hydrogen, Methane, Ethylene, Ethane, Acetylene, and Propane Flames*. In: *Symposium (International) on Combustion*, vol. 24, pp. 129–135. Elsevier.

Gouldin, FC (1996): *Combustion Intensity and Burning Rate Integral of Premixed Flames*. In: *Symposium (International) on Combustion*, vol. 26, pp. 381–388. Elsevier.

Greitzer, EM, Tan, CS and Graf, MB (2004): *Internal Flow: Concepts and Applications*, vol. 3. Cambridge University Press, Cambridge.

Gu, XJ, Haq, MZ, Lawes, M and Woolley, R (2000): *Laminar Burning Velocity and Markstein Lengths of Methane--Air Mixtures*. Combustion and flame, vol. 121, pp. 41–58.

Gülder, OL (1982): *Laminar Burning Velocities of Methanol, Ethanol and Isooctane-Air Mixtures*. In: *Symposium (International) on Combustion*, vol. 19, pp. 275–281. Elsevier.

Gülder, OL (1984): *Correlations of Laminar Combustion Data for Alternative SI Engine Fuels.* Tech. rep., SAE Technical Paper.

Gülder, OL (1991): *Turbulent Premixed Flame Propagation Models for Different Combustion Regimes.* In: *Symposium (International) on Combustion*, vol. 23, pp. 743–750. Elsevier.

Gupta, AK and Lilley, DG (1985): *Flowfield Modeling and Diagnostics.* Energy and engineering science series. Abacus Press, Tunbridge Wells - Kent.

Hall, RS (1961): *Spiral Annular Combustion Chamber.* US Patent 3,000,183.

Hallbäck, M, Henningson, DS, Johansson, AV and Alfredsson, PH (1995): *Turbulence and Transition Modelling.* ERCOFTAC Series. Springer, Berlin.

Hallett, WLH and Toews, DJ (1987): *The Effects of Inlet Conditions and Expansion Ratio on the Onset of Flow Reversal in Swirling Flow in a Sudden Expansion.* Experiments in fluids, vol. 5, pp. 129–133.

Harlow, FH and Nakayama, PI (1967): *Turbulence Transport Equations.* The Physics of Fluids, vol. 10, pp. 2323–2332.

Herrmann, MG (2001): *Numerical Simulation of Premixed Turbulent Combustion Based on a Level Set Flamelet Model.* PhD thesis, Institut für Technische Mechanik, Rheinisch-Westfälischen Technischen Hochschule Aachen. Aachen.

Hinze, JO (1975): *Turbulence.* McGraw-Hill classic textbook reissue series. McGraw-Hill, New York.

Hirsch, C (2007): *Numerical Computation of Internal and External Flows.* Butterworth-Heinemann, Oxford, Oxford, 2nd edn.

HPC (2019): *High Performance Computing (HPC) Und Clustercomputing.* Karlsruhe Institute of Technology (KIT). Online in internet under https://www.scc.kit.edu/dienste/hpc.php (05.05.2019).

Hsieh, TCA, Dahm, WJA and Driscoll, JF (1998): *Scaling Laws for NOx Emission Performance of Burners and Furnaces From 30 KW to 12 MW .* Combustion and Flame, vol. 114, pp. 54–80.

Hsu, KY, Goss, LP, Trump, DD and Roquemore, WM (1995): *Performance of a Trapped-Vortex Combustor.* AIAA paper, p. 810.

Hsu, KY, Goss, LP and Roquemore, WM (1998): *Characteristics of a Trapped-Vortex Combustor.* Journal of Propulsion and Power, vol. 14, pp. 57 65.

Huang, Y and Yang, V (2009): *Dynamics and Stability of Lean-Premixed Swirl-Stabilized Combustion.* Progress in Energy and Combustion Science, vol. 35, pp. 293–364.

ICAO (2016): *Global Air Navigation Plan 2016-2030.* Tech. Rep. Doc 9750-AN/963 Fifth Edition, International Civil Aviation Organization.

Issa, RI (1986): *Solution of the Implicitly Discretised Fluid Flow Equations by Operator-Splitting.* Journal of computational physics, vol. 62, pp. 40–65.

Jasak, H (1996): *Error Analysis and Estimation for the Finite Volume Method with Applications to Fluid Flows.* PhD thesis, Imperial College London (University of London).

Jiménez, J (2004): *Turbulent Flows Over Rough Walls.* Annu. Rev. Fluid Mech, vol. 36, pp. 173–196.

Jiménez, J (2012): *Cascades in Wall-Bounded Turbulence.* Annual Review of Fluid Mechanics, vol. 44, pp. 27–45.

Jochmann, P (2007): *Möglichkeiten Und Grenzen von URANS Verfahren Zur Numerischen Beschreibung Instationärer, Brennkammertypischer Wirbelströmungen.* PhD thesis, Institut für Thermische Strömungsmaschinen, Universität Karlsruhe (TH). Logos Verlag Berlin.

Jochmann, P, Sinigersky, A, Hehle, M, Schäfer, O, Koch, R and Bauer, HJ (2006): *Numerical Simulation of a Precessing Vortex Breakdown.* International Journal of Heat and Fluid Flow, vol. 27, pp. 192–203.

Jones, WP and Launder, BE (1972): *The Prediction of Laminarization with a Two-Equation Model of Turbulence.* International journal of heat and mass transfer, vol. 15, pp. 301–314.

Jouhaud, JC (2010): *Benchmark on the Vortex Preservation.* Tech. rep., CERFACS, Online in internet under http://elearning.cerfacs.fr/numerical/benchmarks/vortex2d/index.php (05.05.2019).

Kao, YH, Tambe, SB and Jeng, SM (2014): *Aerodynamics Study of a Linearly-Arranged 5-Swirler Array.* In: *ASME Turbo Expo 2014: Turbine Technical Conference and Exposition,* pp. V04AT04A007,1–V04AT04A007,12. American Society of Mechanical Engineers.

Kärrholm, FB, Tao, F and Nordin, N (2008): *Three-Dimensional Simulation of Diesel Spray Ignition and Flame Lift-Off Using OpenFOAM and KIVA-3V CFD Codes.* In: *SAE Technical Paper,* pp. 1–18. SAE International.

Kestin, J and Leidenfrost, W (1959): *An Absolute Determination of the Viscosity of Eleven Gases Over a Range of Pressures.* Physica, vol. 25, pp. 1033–1062.

Kim, J, Moin, P and Moser, R (1987): *Turbulence Statistics in Fully Developed Channel Flow at Low Reynolds Number.* Journal of fluid mechanics, vol. 177, pp. 133–166.

Kim, SW (1991): *Calculation of Divergent Channel Flows with a Multiple-Time-Scale Turbulence Model.* AIAA journal, vol. 29, pp. 547–554.

Kobayashi, H, Tamura, T, Maruta, K, Niioka, T and Williams, FA (1996): *Burning Velocity of Turbulent Premixed Flames in a High-Pressure Environment.* Symposium (International) on Combustion, vol. 26, pp. 389–396.

Kobayashi, H, Nakashima, T, Tamura, T, Maruta, K and Niioka, T (1997): *Turbulence Measurements and Observations of Turbulent Premixed Flames at Elevated Pressures up to 3.0 MPa.* Combustion and Flame, vol. 108, pp. 104–117.

Kobayashi, H, Kawabata, Y and Maruta, K (1998): *Experimental Study on General Correlation of Turbulent Burning Velocity at High Pressure.* Symposium (International) on Combustion, vol. 27, pp. 941–948. Twenty-Seventh Sysposium (International) on Combustion Volume One.

Kolmogorov, AN (1941): *The Local Structure of Turbulence in Incompressible Viscous Fluid for Very Large Reynolds Numbers.* Dokl. Akad. Nauk SSSR, vol. 30, pp. 299–303.

Kolmogorov, AN (1962): *A Refinement of Previous Hypotheses Concerning the Local Structure of Turbulence in a Viscous Incompressible Fluid at High Reynolds Number.* Journal of Fluid Mechanics, vol. 13, pp. 82–85.

Kolmogorov, AN, Petrovskii, IG and Piskunov, NS (1937): *A Study of the Equation of Diffusion with Increase in the Quantity of Matter, and Its Application to a Biological Problem.* Bjul. Moskovskogo Gos. University, vol. 1, pp. 1–26.

Kröner, M, Sattelmayer, T, Fritz, J, Kiesewetter, F and Hirsch, C (2007): *Flame Propagation in Swirling Flows-Effect of Local Extinction on the Combustion Induced Vortex Breakdown.* Combustion Science and Technology, vol. 179, pp. 1385–1416.

Launder, BE and Sharma, BI (1974): *Application of the Energy-Dissipation Model of Turbulence to the Calculation of Flow Near a Spinning Disc.* Letters in heat and mass transfer, vol. 1, pp. 131–137.

Launder, BE and Spalding, DB (1974): *The Numerical Computation of Turbulent Flows.* Computer methods in applied mechanics and engineering, vol. 3, pp. 269–289.

Launder, BE, Reece, GJ and Rodi, W (1975): *Progress in the Development of a Reynolds-Stress Turbulence Closure.* Journal of fluid mechanics, vol. 68, pp. 537–566.

Laverdant, AM and Candel, SM (1989): *Computation of Diffusion and Premixed Flames Rolled up in Vortex Structures.* vol. 5, pp. 134–143.

Law, CK (1993): *A Compilation of Recent Experimental Data of Premixed Laminar Flames.* Reduced Kinetic Mechanisms for Applications in Combustion Systems, Lecture Notes in Physics, vol. 15, pp. 19–30.

Law, CK (2010): *Combustion Physics.* Cambridge University Press, Cambridge.

Le, H, Moin, P and Kim, J (1997): *Direct Numerical Simulation of Turbulent Flow Over a Backward-Facing Step.* Journal of fluid mechanics, vol. 330, pp. 349–374.

Lecocq, G, Richard, S, Colin, O and Vervisch, L (2011): *Hybrid Presumed Pdf and Flame Surface Density Approaches for Large-Eddy Simulation of Premixed Turbulent Combustion: Part 1: Formalism and Simulation of a Quasi-Steady Burner*. Combustion and Flame, vol. 158, pp. 1201–1214.

Lee, M and Moser, RD (2015): *Direct Numerical Simulation of Turbulent Channel Flow up to $Re_\tau \approx 5200$*. Journal of Fluid Mechanics, vol. 774, pp. 395–415.

Lefebvre, AH and Ballal, DR (2010): *Gas Turbine Combustion: Alternative Fuels and Emissions, Third Edition*. Taylor & Francis, New York.

Leibovich, S (1978): *The Structure of Vortex Breakdown*. Annual review of fluid mechanics, vol. 10, pp. 221–246.

Leibovich, S (1984): *Vortex Stability and Breakdown- Survey and Extension*. AIAA journal, vol. 22, pp. 1192–1206.

Leschziner, MA and Drikakis, D (2002): *Turbulence Modelling and Turbulent-Flow Computation in Aeronautics*. The Aeronautical Journal, vol. 106, pp. 349–384.

Libby, PA, Bilger, RW and Williams, FA (1980): *Turbulent Reacting Flows*. Topics in applied physics. Springer, Berlin.

Lipatnikov, AN and Chomiak, J (2002): *Turbulent Flame Speed and Thickness: Phenomenology, Evaluation, and Application in Multi-Dimensional Simulations*. Progress in Energy and Combustion Science, vol. 28, pp. 1–74.

Livesey, JL and Hugh, T (1966): *Suitable Mean Values in One-Dimensional Gas Dynamics*. Journal of Mechanical Engineering Science, vol. 8, pp. 374–383.

Lucca-Negro, O and O'doherty, T (2001): *Vortex Breakdown: a Review*. Progress in Energy and Combustion Science, vol. 27, pp. 431–481.

Magnussen, BF and Hjertager, BW (1981): *On the Structure of Turbulence and a Generalized Eddy Dissipation Concept for Chemical Reaction in Turbulent Flow*. In: *19th AIAA Aerospace Meeting, St. Louis, USA*, vol. 198, p. 42.

Mancini, AA, Burrus, DL and Lohmueller, SJ (2007): *Method and Apparatus for Assembling Gas Turbine Engine*. EP Patent App. EP20,070,103,306.

Mauß, F and Peters, N (1993): *Reduced Kinetic Mechanisms for Premixed Methane-Air Flames*. Reduced kinetic mechanisms for applications in combustion systems, pp. 58–75.

Meier, W, Weigand, P, Duan, XR and Giezendanner-Thoben, R (2007): *Detailed Characterization of the Dynamics of Thermoacoustic Pulsations in a Lean Premixed Swirl Flame*. Combustion and Flame, vol. 150, pp. 2–26.

Menter, FR (1994): *Two-Equation Eddy-Viscosity Turbulence Models for Engineering Applications*. AIAA journal, vol. 32, pp. 1598–1605.

Metghalchi, M and Keck, JC (1980): *Laminar Burning Velocity of Propane-Air Mixtures at High Temperature and Pressure*. Combustion and Flame, vol. 38, pp. 143–154.

Moin, P, Squires, K, Cabot, W and Lee, S (1991): *A Dynamic Subgrid-Scale Model for Compressible Turbulence and Scalar Transport*. Physics of Fluids A: Fluid Dynamics, vol. 3, pp. 2746–2757.

Moureau, V, Lartigue, G, Sommerer, Y, Angelberger, C, Colin, O and Poinsot, T (2005): *Numerical Methods for Unsteady Compressible Multi-Component Reacting Flows on Fixed and Moving Grids*. Journal of Computational Physics, vol. 202, pp. 710–736.

Moureau, V, Minot, P, Pitsch, H and Bérat, C (2007): *A Ghost-Fluid Method for Large-Eddy Simulations of Premixed Combustion in Complex Geometries*. Journal of Computational Physics, vol. 221, pp. 600–614.

Mydlarski, L and Warhaft, Z (1998): *Passive Scalar Statistics in High-Péclet-Number Grid Turbulence*. Journal of Fluid Mechanics, vol. 358, pp. 135–175.

Negulescu, DA (2013): *Gas Turbine Combustion Chamber Arrangement of Axial Type of Construction*. US Patent App. 13/704,972.

Negulescu, DA (2014): *Gas Turbine Centripetal Annular Combustion Chamber and Method for Flow Guidance*. US Patent App. 14/232,814.

Nicoud, F and Ducros, F (1999): *Subgrid-Scale Stress Modelling Based on the Square of the Velocity Gradient Tensor*. Flow, turbulence and Combustion, vol. 62, pp. 183–200.

Nicoud, F, Toda, HB, Cabrit, O, Bose, S and Lee, J (2011): *Using Singular Values to Build a Subgrid-Scale Model for Large Eddy Simulations*. Physics of Fluids, vol. 23, pp. 085106,1–085106,12.

OpenCFD (2015): *OpenFOAM 2.4.0 -The Open Source CFD Toolbox User's Guide*. OpenCFD Ltd.

Orszag, SA, Staroselsky, I, Flannery, WS and Zhang, Y (1996): *Introduction to Renormalization Group Modeling of Turbulence*. Simulation and Modeling of Turbulent Flows, pp. 155–184.

Peters, N (1991): *Length Scales in Laminar and Turbulent Flames*. Progress in Astronautics and Aeronautics, vol. 135, pp. 155–182.

Peters, N (1997): *Kinetic Foundation of Thermal Flame Theory*. Advances in Combustion Science: In honor of Ya. B. Zel'dovich(A 97-24531 05-25), Reston, VA, American Institute of Aeronautics and Astronautics, Inc.(Progress in Astronautics and Aeronautics, vol. 173, pp. 73–91.

Peters, N (1999): *The Turbulent Burning Velocity for Large-Scale and Small-Scale Turbulence*. Journal of Fluid mechanics, vol. 384, pp. 107–132.

Peters, N (2000): *Turbulent Combustion*. Cambridge Monographs on Mechanics. Cambridge University Press, Cambridge.

Peters, N and Donnerhack, S (1981): *Structure and Similarity of Nitric Oxide Production in Turbulent Diffusion Flames*. Symposium (International) on Combustion, vol. 18, pp. 33–42. Eighteenth Symposium (International) on Combustion.

Peters, N and Williams, FA (1987): *The Asymptotic Structure of Stoichiometric Methane-Air Flames*. Combustion and Flame, vol. 68, pp. 185–207.

Philip, M, Boileau, M, Vicquelin, R, Schmitt, T, Durox, D, Bourgouin, JF and Candel, S (2015): *Simulation of the Ignition Process in an Annular Multiple-Injector Combustor and Comparison with Experiments*. Journal of Engineering for Gas Turbines and Power, vol. 137, pp. 031501,1–031501,9.

Pianko, M and Wazelt, F (1983): *Suitable Averaging Techniques in Non-Uniform Internal Flows. Propulsion and Energetics Panel Working Group 14 on Suitable Averaging Techniques in Non-Uniform Internal Flows*. AGARD Advisory Report.

Pitsch, H (2006): *Large-Eddy Simulation of Turbulent Combustion*. Annu. Rev. Fluid Mech, vol. 38, pp. 453–482.

Poinsot, T and Lelef, SK (1992): *Boundary Conditions for Direct Simulations of Compressible Viscous Flows*. Journal of Computational Physics, vol. 101, pp. 104–129.

Poinsot, T and Veynante, D (2012): *Theoretical and Numerical Combustion*. R.T. Edwards, Philadelphia, 3rd edn.

Poinsot, T, Veynante, D and Candel, S (1991): *Diagrams of Premixed Turbulent Combustion Based on Direct Simulation*. Symposium (International) on Combustion, vol. 23, pp. 613–619. Twenty-Third Symposium (International) on Combustion.

Pope, SB (1975): *A More General Effective-Viscosity Hypothesis*. Journal of Fluid Mechanics, vol. 72, pp. 331–340.

Pope, SB (2000): *Turbulent Flows*. Cambridge University Press, Cambridge.

Prandtl, L (1925): *Über die Ausgebildete Turbulenz (Investigations on Turbulent Flow)*. Z. Angew. Math. Mech, vol. 5, pp. 136–139.

Prandtl, L (1932): *Zur Turbulenten Strömung in Rohren Und Längs Platten*. Ergebnisse der aerodynamischen Versuchsanstalt zu Göttingen, vol. 4, pp. 18–29.

Reba, I (1966): *Applications of The Coanda Effect*. Scientific American, vol. 214, pp. 84–93.

Renard, PH, Thevenin, D, Rolon, JC and Candel, S (2000): *Dynamics of Flame/vortex Interactions*. Progress in Energy and Combustion Science, vol. 26, pp. 225–282.

Reynolds, O (1894): *On the Dynamical Theory of Incompressible Viscous Fluids and the Determination of the Criterion*. Proceedings of the Royal Society of London, vol. 56, pp. 40–45.

Reynolds, WC (1990): *The Potential and Limitations of Direct and Large Eddy Simulations*, vol. 357 of *Lecture Notes in Physics*, pp. 313–343. Springer, Berlin, Heidelberg.

Rogallo, RS and Moin, P (1984): *Numerical Simulation of Turbulent Flows*. Annual review of fluid mechanics, vol. 16, pp. 99–137.

Roquemore, WM, Shouse, D, Burrus, D, Johnson, A, Cooper, C, Duncan, B, Hsu, KY, Katta, V, Sturgess, G and Vihinen, I (2001): *Vortex Combustor Concept for Gas Turbine Engines*. In: *39th Aerospace Sciences Meeting and Exhibit*, pp. 8–11.

Roux, S, Lartigue, G, Poinsot, T, Meier, W and Bérat, C (2005): *Studies of Mean and Unsteady Flow in a Swirled Combustor Using Experiments, Acoustic Analysis, and Large Eddy Simulations*. Combustion and Flame, vol. 141, pp. 40–54.

Rusak, Z, Kapila, AK and Choi, JJ (2002): *Effect of Combustion on Near-Critical Swirling Flow*. Combustion Theory and Modelling, vol. 6, pp. 625–645.

Samarasinghe, J, Peluso, S, Szedlmayer, M, De Rosa, A, Quay, B and Santavicca, D (2013): *Three-Dimensional Chemiluminescence Imaging of Unforced and Forced Swirl-Stabilized Flames in a Lean Premixed Multi-Nozzle Can Combustor*. Journal of Engineering for Gas Turbines and Power, vol. 135, pp. 101503.

Schmid, HP (1995): *Ein Verbrennungsmodell Zur Beschreibung der Wärmefreisetzung von Vorgemischten Turbulenten Flammen*. PhD thesis, Engler Bunte Institute, Karlsruhe Institute of Technology. Karlsruhe.

Schmid, HP, Habisreuther, P and Leuckel, W (1998): *A Model for Calculating Heat Release in Premixed Turbulent Flames*. Combustion and Flame, vol. 113, pp. 79–91.

Schobeiri, MT (2010): *Fluid Mechanics for Engineers: A Graduate Textbook*. Springer, Berlin.

Schutz, H, Kraupa, W and Termuhlen, H (1999): *Gas Turbine Engine with Tilted Burners*. US Patent 5,946,902.

Seippel, C (1943): *Gas Turbine Plant*. US Patent 2,326,072.

Selle, L, Lartigue, G, Poinsot, T, Koch, R, Schildmacher, KU, Krebs, W, Prade, B, Kaufmann, P and Veynante, D (2004): *Compressible Large Eddy Simulation of Turbulent Combustion in Complex Geometry on Unstructured Meshes*. Combustion and Flame, vol. 137, pp. 489–505.

Seshadri, K and Peters, N (1990): *The Inner Structure of Methane-Air Flames*. Combustion and Flame, vol. 81, pp. 96–118.

Shih, TH, Liou, WW, Shabbir, A, Yang, Z and Zhu, J (1995): *A New K-ε Eddy Viscosity Model for High Reynolds Number Turbulent Flows*. Computers & Fluids, vol. 24, pp. 227–238.

Smagorinsky, J (1963): *General Circulation Experiments with the Primitive Equations: I. The Basic Experiment*. Monthly weather review, vol. 91, pp. 99–164.

Smart, JP, Morgan, DJ and Roberts, PA (1992): *The Effect of Scale on the Performance of Swirl Stabilised Pulverised Coal Burners*. Symposium (International) on Combustion, vol. 24, pp. 1365–1372. Twenty-Fourth Symposium on Combustion.

Smith, GP, Golden, DM, Frenklach, M, Moriarty, NW, Eiteneer, B, Goldenberg, M, Bowman, CT, Hanson, RK, Song, S and Gardiner, WC (2011): *GRI-Mech 3.0, 1999*. Online in internet under http://combustion.berkeley.edu/gri-mech/version30/text30.html (05.05.2019).

Smith, LM and Reynolds, WC (1992): *On the Yakhot-Orszag Renormalization Group Method for Deriving Turbulence Statistics and Models*. Physics of Fluids A: Fluid Dynamics, vol. 4, pp. 364–390.

Smith, LM and Woodruff, SL (1998): *Renormalization-Group Analysis of Turbulence*. Annual Review of Fluid Mechanics, vol. 30, pp. 275–310.

Spalding, DB (1971): *Mixing and Chemical Reaction in Steady Confined Turbulent Flames*. In: *Symposium (International) on Combustion*, vol. 13, pp. 649–657. Elsevier.

Spalding, DB, Hottel, HC, Bragg, SL, Lefebvre, AH, Shepherd, DG and Scurlock, AC (1963): *The Art of Partial Modeling*. Symposium (International) on Combustion, vol. 9, pp. 833–843.

Speziale, CG, Sarkar, S and Gatski, TB (1991): *Modelling the Pressure--Strain Correlation of Turbulence: an Invariant Dynamical Systems Approach*. Journal of fluid mechanics, vol. 227, pp. 245–272.

Squire, HB (1960): *Analysis of the Vortex Breakdown Phenomenon*. Imperial College of Science and Technology, London.

Staffelbach, G, Gicquel, LYM, Boudier, G and Poinsot, T (2009): *Large Eddy Simulation of Self Excited Azimuthal Modes in Annular Combustors*. Proceedings of the Combustion Institute, vol. 32, pp. 2909–2916.

Stein, O and Kempf, A (2007): *LES of the Sydney Swirl Flame Series: A Study of Vortex Breakdown in Isothermal and Reacting Flows*. Proceedings of the Combustion Institute, vol. 31, pp. 1755 – 1763.

Strokin, VN, Shikhman, YM and VP, Ljashenko (2013): *Low Emission, Compact Combustor With Air Flow Swirling at the Entrance*. International Society for Air Breathing Engines (ISABE), vol. 1527, pp. 1305–1314.

Strokin, VN, Volkov, SA, Ljashenko, VP, Popov, VI, Startzev, AN, Nigmatullin, RZ, Shilova, TV and Belikov, UV (2017): *Compact Combustor Integrated (CI) with Compressor and Turbine for Perspective Turbojet Engine*. Journal of Physics: Conference Series, vol. 891, pp. 012241,1–012241,6.

Sutherland, W (1893): *The Viscosity of Gases and Molecular Force*. The London, Edinburgh, and Dublin Philosophical Magazine and Journal of Science, vol. 36, pp. 507–531.

Syred, N (2006): *A Review of Oscillation Mechanisms and the Role of the Precessing Vortex Core (PVC) in Swirl Combustion Systems*. Progress in Energy and Combustion Science, vol. 32, pp. 93–161.

Syred, N and Beer, JM (1974): *Combustion in Swirling Flows: a Review*. Combustion and flame, vol. 23, pp. 143–201.

Syred, N, Gupta, AK and Beer, JM (1975): *Temperature and Density Gradient Changes Arising with the Precessing Vortex Core and Vortex Breakdown in Swirl Burners*. In: *Symposium (International) on Combustion*, vol. 15, pp. 587–597. Elsevier.

Syred, N, Fick, W, O'doherty, T and Griffiths, AJ (1997): *The Effect of the Precessing Vortex Core on Combustion in a Swirl Burner*. Combustion science and technology, vol. 125, pp. 139–157.

Tangirala, VE and Joshi, ND (2015): *Ultra Compact Combustor*. US Patent App. 14/706,679.

Traupel, W (1978): *Thermal Turbomachines*, vol. 1. Springer, Berlin, 3rd edn.

Trouvé, A and Poinsot, T (1994): *The Evolution Equation for the Flame Surface Density in Turbulent Premixed Combustion*. Journal of Fluid Mechanics, vol. 278, pp. 1–31.

Turns, SR (2012): *An Introduction to Combustion: Concepts and Applications*. McGraw-Hill - New York.

Tyler, RD (1957): *One-Dimensional Treatment of Non-Uniform Flow*. Tech. Rep. 2991, Ministry of Supply, Aeronautical Research Council Reports and Memoranda, Her Majesty's Stationery Office, London.

van Driest, ER (1956): *On Turbulent Flow Near a Wall*. J. Aeronaut. Sci, vol. 23, pp. 1007–1011.

von Kármán, T (1930): *Mechanische Ähnlichkeit Und Turbulenz*. Nachrichten von der Gesellschaft der Wissenschaften zu Göttingen, Mathematisch-physische Klasse, vol. 5, pp. 58–76.

Wallin, S and Johansson, AV (2000): *An Explicit Algebraic Reynolds Stress Model for Incompressible and Compressible Turbulent Flows*. Journal of Fluid Mechanics, vol. 403, pp. 89–132.

Wang, P, Platova, NA, Fröhlich, J and Maas, U (2014): *Large Eddy Simulation of the PRECCIN-STA Burner*. International Journal of Heat and Mass Transfer, vol. 70, pp. 486–495.

Wang, P, Froehlich, J, Maas, U, He, ZX and Wang, CJ (2016): *A Detailed Comparison of Two Sub-Grid Scale Combustion Models Via Large Eddy Simulation of the PRECCINSTA Gas Turbine Model Combustor*. Combustion and Flame, vol. 164, pp. 329–345.

Weber, R (1996): *Scaling Characteristics of Aerodynamics, Heat Transfer, and Pollutant Emissions in Industrial Flames.* Symposium (International) on Combustion, vol. 26, pp. 3343–3354.

Weber, R and Breussin, F (1998): *Scaling Properties of Swirling Pulverized Coal Flames: From 180 KW to 50 MW Thermal Input.* Symposium (International) on Combustion, vol. 27, pp. 2957–2964.

Weigand, P, Duan, XR, Meier, W, Meier, U, Aigner, M and Bérat, C (2005): *Experimental Investigations of an Oscillating Lean Premixed CH4-Air Swirl Flame in a Gas Turbine Model Combustor.* In: *Proc. of the European Combustion Meeting 2005*, pp. 235–240.

Weigand, P, Meier, W, Duan, X and Aigner, M (2007): *Laser-Based Investigations of Thermoacoustic Instabilities in a Lean Premixed Gas Turbine Model Combustor.* Journal of Engineering for Gas Turbines and power, vol. 129, pp. 664–671.

Weller, HG, Tabor, G, Jasak, H and Fureby, C (1998): *A Tensorial Approach to Computational Continuum Mechanics Using Object-Oriented Techniques.* Computers in physics, vol. 12, pp. 620–631.

Werschnik, H, Hilgert, J, Wilhelm, M, Bruschewski, M and Schiffer, HP (2017a): *Influence of Combustor Swirl on Endwall Heat Transfer and Film Cooling Effectiveness at the Large Scale Turbine Rig.* Journal of Turbomachinery, vol. 139, pp. 081007,1–081007,12.

Werschnik, H, Schneider, M, Herrmann, J, Ivanov, D, Schiffer, HP and Lyko, C (2017b): *The Influence of Combustor Swirl on Pressure Losses and the Propagation of Coolant Flows at the Large Scale Turbine Rig (LSTR): Experimental and Numerical Investigation.* International Journal of Turbomachinery, Propulsion and Power, vol. 2, pp. 1–18.

Westbrook, CK and Dryer, FL (1981): *Simplified Reaction Mechanisms for the Oxidation of Hydrocarbon Fuels in Flames.* Combustion science and technology, vol. 27, pp. 31–43.

Wilcox, DC (1988): *Multiscale Model for Turbulent Flows.* AIAA journal, vol. 26, pp. 1311–1320.

Wilcox, DC (2006): *Turbulence Modeling for CFD.* No. v. 1 in Turbulence Modeling for CFD. DCW Industries, Sherman Oaks-California, 3rd edn.

Wille, R and Fernholz, H (1965): *Report on the First European Mechanics Colloquium, on the Coanda Effect.* Journal of Fluid Mechanics, vol. 23, pp. 801–819.

Williams, FA (1985): *Combustion Theory: The Fundamental Theory of Chemically Reacting Flow Systems.* Benjamin/Cummings, Menlo Park - California, 2nd edn.

Wyatt, DD (1955): *Analysis of Errors Introduced by Several Methods of Weighting Nonuniform Duct Flows.*

Yakhot, V, Orszag, SA, Thangam, S, Gatski, TB and Speziale, CG (1992): *Development of Turbulence Models for Shear Flows by a Double Expansion Technique.* Physics of Fluids, vol. 4, pp. 1510–1520.

Zel'dovich, YB and Frank-Kamenetzki, DA (1938): *A Theory of Thermal Propagation of Flame.* Acta Physiochimica URSS IX, pp. 341–350.

Zelina, J, Ehret, J, Hancock, RD, Shouse, DT, Sturgess, GJ and Roquemore, WM (2002): *Ultra-Compact Combustion Technology Using High Swirl for Enhanced Burning Rate.* AIAA Paper, p. 3725.

Zelina, J, Sturgess, GJ and Shouse, DT (2004): *The Behavior of an Ultra-Compact Combustor (UCC) Based on Centrifugally-Enhanced Turbulent Burning Rates.* AIAA Paper, p. 3541.

Zhang, F (2014): *Numerical Modeling of Noise Generated by Turbulent Combustion.* PhD thesis, Engler Bunte Institute, Karlsruhe Institute of Technology. Shaker Verlag GmbH, Aachen.

Zhang, F, Habisreuther, P, Hettel, M and Bockhorn, H (2009): *Modelling of a Premixed Swirl-Stabilized Flame Using a Turbulent Flame Speed Closure Model in LES.* Flow, Turbulence and Combustion, vol. 82, pp. 537–551.

Zhang, F, Habisreuther, P and Bockhorn, H (2013): *Application of the Unified Turbulent Flame-Speed Closure (UTFC) Combustion Model to Numerical Computation of Turbulent Gas Flames.* High Performance Computing in Science and Engineering, pp. 187–205.

Zhang, J, Nieh, S and Zhou, L (1992): *A New Version of Algebraic Stress Model for Simulating Strongly Swirling Turbulent Flows.* Numerical Heat Transfer, Part B Fundamentals, vol. 22, pp. 49–63.

Zhang, Y and Orszag, SA (1998): *Two-Equation RNG Transport Modeling of High Reynolds Number Pipe Flow.* Journal of scientific computing, vol. 13, pp. 471–483.

Zimont, VL and Lipatnikov, AN (1995): *A Numerical Model of Premixed Turbulent Combustion of Gases.* Chem. Phys. Rep, vol. 14, pp. 993–1025.

Zucker, RD and Biblarz, O (2002): *Fundamentals of Gas Dynamics.* John Wiley & Sons, Hoboken - New Jersey.

Student Theses

Holz, S (2015): *Grobstruktursimulation der Strömung in einer neuartigen Gasturbinen-Brennkammer.* Master's thesis, Supervisor: Bauer, HJ, Co-supervisor: Ariatabar, B, Institute of Thermal Turbomachinery, Karlsruhe Institute of Technology.

Klatt, JN (2012): *Numerische Untersuchung einer Brennkammerströmung mittels URANS Verfahren.* Student thesis, Supervisor: Bauer, HJ, Co-supervisor: Ariatabar, B, Institute of Thermal Turbomachinery, Karlsruhe Institute of Technology.

Schäfer, F (2018): *Numerical Analysis of Asymmetrically Bounded Swirling Flows.* Master's thesis, Supervisor: Bauer, HJ, Co-supervisor: Ariatabar, B, Institute of Thermal Turbomachinery, Karlsruhe Institute of Technology.

Schneider, MS (2014): *Large Eddy Simulation of Isothermal and Reacting Flow in a Gas Turbine Model Combustor.* Master's thesis, Supervisor: Bauer, HJ, Co-supervisor: Ariatabar, B, Institute of Thermal Turbomachinery, Karlsruhe Institute of Technology.

Trapp, S (2014): *Implementierung und Validierung eines "Turbulent Flame Speed Closure" Verbrennungsmodells zur Simulation der Drallflammen.* Master's thesis, Supervisor: Bauer, HJ, Co-supervisor: Ariatabar, B, Institute of Thermal Turbomachinery, Karlsruhe Institute of Technology.

Wilhelm, S (2013): *Numerical investigation of flow phenomena in a novel gas turbine combustion system concept.* Master's thesis, Supervisor: Bauer, HJ, Co-supervisor: Ariatabar, B, Institute of Thermal Turbomachinery, Karlsruhe Institute of Technology.

List of Figures

Appendix

A.1 Direct Numerical, Large Eddy and Reynolds Averaged Simulations

The governing equations introduced in Sec. 2.1.1 describe every detail of the fluid motion. The DNS has as objective to numerically solve these equations over the complete spatiotemporal range of the turbulent scales (Kim et al., 1987, Le et al., 1997, Lee and Moser, 2015, Rogallo and Moin, 1984). The formidable challenge is thereby, the vast computation power and memory required to capture the entire physics of the flow.

An example of homogeneous isotropic turbulence should elucidate the problem. The size of Kolmogorov eddies in such flows is inversely proportional to $Re^{3/4}$ (Reynolds, 1990). Resolving the smallest eddy by N grid points per unit length, the required memory to handle the mesh, as well as the number of arithmetic operations, will scale with $N^3 \sim Re^{9/4}$. Moreover, the governing equations must be integrated in time to capture the transient nature of turbulence. The required number of time steps is determined by the Kolmogorov time scale, which is proportional to $M \sim Re^{3/4}$ (Pope, 2000). The total computational expense is then proportional to the number of operations times the number of steps $N^3 \cdot M \sim Re^3$, and that only just for homogeneous isotropic turbulence.

Turbulent flows in nature are prevalently inhomogeneous and anisotropic. The local grid resolution for DNS of such cases needs to be additionally adapted, for example in the vicinity of walls (Jiménez, 2004, 2012). Hence, the technical flows with Reynolds numbers in order of 10^5-10^7 put such a high demand on computational capacity, rendering this method inappropriate for engineering applications.

In LES as a remedy for DNS, the governing equations are being solved only for large energy containing and prevalently anisotropic turbulent eddies. The small scales, on the other hand, can be modeled with straightforward semi-empirical laws because they are inherently homogeneous and isotropic. A spatial filter, which is often the grid size, separate the large scale structures from the small sub-grid scales (SGS). Increasing the mesh resolution, the scales to be modeled get smaller, and thus the solution approaches the DNS.

Despite the filtering technique and modeling of the computationally expensive small scales, the LES cannot be the reasonable choice for many practical problems. Analogous to DNS, it can be shown that the total computational effort for LES scales with $Re^{9/4}$, which is significantly lower than the DNS (Hirsch, 2007). However, depending on the application and the nature of the flow, the excessive computational expense for LES still would not be reasonable, while the flow can be predicted via statistical approaches at a sufficient level of description with acceptable accuracy and much lower cost.

Prediction of the combustion phenomena is a domain, where LES gains increasing importance also in practical applications. A review of LES of turbulent combustion can be found in Pitsch (2006) as well as in the textbook of Echekki and Mastorakos (2010). In Chap. 3, a benchmarking of LES and RANS regarding the requirements of this work is performed, and the decision to use

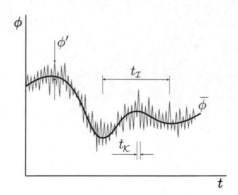

Figure A.1: Reynolds decomposition and ensemble averaging of a turbulent flow variable.

RANS is discussed.

RANS is a rational alternative to DNS and LES pursuing the modeling of turbulence via a statistical approach. Accordingly, the instantaneous flow variable ϕ is decomposed into a mean value $\bar{\phi}$ and the fluctuation around that ϕ' as proposed by Reynolds (1894)

$$\phi(\boldsymbol{r},t) = \bar{\phi}(\boldsymbol{r},t) + \phi'(\boldsymbol{r},t) \tag{A.1}$$

According to Hinze (1975), three main averaging methods can be employed to statistically model the turbulent flows based on the Reynolds decomposition. These are the "time averaging" for stationary turbulence, the "spatial averaging" for homogeneous turbulence and the "ensemble averaging" as the most general form suitable for inhomogeneous and unsteady turbulence. For N repeatable, independent and identically distributed realization of a turbulent flow property, the ensemble average of the variable $\phi(\boldsymbol{r},t)$ is (Pope, 2000)

$$\bar{\phi}(\boldsymbol{r},t) = \frac{1}{N}\sum_{i=1}^{N}\phi_i(\boldsymbol{r},t) \tag{A.2}$$

The N members of the ensemble are numerically generated by varying the initial condition for each realization with the aid of a normal-distribution random-number generator. Thereby, all other controllable variables (such as energy spectrum, shear rate, boundary condition) are identical (Ferziger and Peric, 2012).

The turbulent flow variable $\phi(\boldsymbol{r},t)$ and the corresponding time scales are schematically illustrated in Fig. A.1. The length of the time step of simulation Δt as well as the number of members of the ensemble N per time step must be large enough to fade the effects of the small scale turbulent fluctuations with the time scale t_κ

$$\overline{\phi'}(\boldsymbol{r},t) = 0 \tag{A.3}$$

On the other hand, Δt must be far smaller than the time scale of the energy containing eddies $t_\mathcal{I}$

$$t_\kappa \ll \Delta t \ll t_\mathcal{I} \tag{A.4}$$

Inequation (A.4) presumes implicitly that $t_\kappa \ll t_\mathcal{I}$, which is a justified assumption for most of transient flows of engineering interest (Wilcox, 2006).

In the case of significant density changes in the flow, such that occur in combustion, it is practical to use a density-weighted decomposition of instantaneous variable as introduced by Favre (1965)

$$\phi(r,t) = \tilde{\phi}(r,t) + \phi''(r,t) \quad \text{with} \quad \overline{\rho\phi''} = 0 \tag{A.5}$$

The definition for "Favre averaged" $\tilde{\phi}$ can be then derived by multiplying the density ρ with Eq. (A.5) and subsequently averaging

$$\tilde{\phi} = \frac{\overline{\rho\phi}}{\bar{\rho}} \tag{A.6}$$

With this mathematical simplification, the density fluctuation can be eliminated from the averaged governing equations, however, their physical effect on the turbulence do not disappear (Libby et al., 1980).

With Eqs. (A.2) and (A.5), the governing equations can be directly derived for time-dependent mean properties. They are generally known as the Unsteady Reynolds Averaged Navier–Stokes equations (URANS). The averaging of the nonlinear convective term in the Navier–Stokes equations gives rise to appearance of momentum fluxes, which act naturally similar to viscous shear stresses within the fluid

$$\frac{\partial \bar{\rho}\tilde{u}}{\partial t} + \nabla \cdot (\bar{\rho}\tilde{u}\tilde{u} + \bar{\rho}\widetilde{u''u''}) = \bar{\rho}f_v - \nabla\bar{p} + \nabla\cdot\tilde{\tau} \tag{A.7}$$

The corresponding correlation $\bar{\rho}\widetilde{u''u''}$ is apriori unknown and called Reynolds stress tensor. The subject of turbulence modeling in RANS framework is the treatment of the Reynolds stress tensor, known as the "closure problem".

In the recent decades, a broad spectrum of turbulence models with different levels of complexity has been developed and tested. It spreads from the pioneering algebraic model of Ludwig Prandtl based on his mixing length hypothesis (Prandtl, 1925), to the derivation of transport equations for the six components of the Reynolds stress tensor (Launder et al., 1975). However, due to the irregular nature of turbulence, none of these models has been able to represent all types and scales of turbulent structures universally. Therefore, the usual approach is to choose the turbulence model depending on the specific flow problem and calibrate it via several model constants. A comprehensive review of RANS turbulence models and their field of applications can be found in Leschziner and Drikakis (2002) and Wilcox (2006).

A widely accepted approach for modeling the Reynolds stresses is the eddy viscosity approximation of Boussinesq (1877). It assumes overall an isotropic turbulence and prescribes, analogous to viscous shear stresses (cf. 2.8), a "linear relation" between the Reynolds stresses and the mean velocity gradient

$$\bar{\rho}\widetilde{u''u''} = -\mu_t \left[\nabla\tilde{u} + (\nabla\tilde{u})^T - \frac{2}{3}(\nabla\cdot\tilde{u})I \right] + \frac{2}{3}\bar{\rho}\tilde{k}I \quad \text{with} \quad \tilde{k} = \frac{1}{2}\widetilde{u''\cdot u''} \tag{A.8}$$

The standard k,ε-model is the most widely used approach to evaluate the eddy viscosity μ_t, proposed and improved by Harlow and Nakayama (1967), Jones and Launder (1972), Launder and Sharma (1974). It is defined by the turbulent kinetic energy k and its dissipation rate ε as

$$\mu_t = \rho\, c_\mu \frac{k^2}{\varepsilon} \quad \text{with} \quad c_\mu = 0.09 \tag{A.9}$$

where the Favre averaged variables $\tilde{k}$ and $\tilde{\varepsilon}$ are described by a closure of their transport equations in most general form as (Libby et al., 1980)

$$\bar{\rho}\frac{\partial\tilde{k}}{\partial t} + \bar{\rho}\tilde{u}\cdot\nabla\tilde{k} = \nabla\cdot\left(\frac{\mu_t}{\sigma_k}\nabla\tilde{k}\right) - \bar{\rho}\widetilde{u''u''}:\nabla\tilde{u} - \bar{\rho}\tilde{\varepsilon} \tag{A.10}$$

$$\bar{\rho}\frac{\partial\tilde{\varepsilon}}{\partial t} + \bar{\rho}\tilde{u}\cdot\nabla\tilde{\varepsilon} = \nabla\cdot\left(\frac{\mu_t}{\sigma_\varepsilon}\nabla\tilde{\varepsilon}\right) - c_{\varepsilon1}\frac{\tilde{\varepsilon}}{\tilde{k}}\bar{\rho}\widetilde{u''u''}:\nabla\tilde{u} - c_{\varepsilon2}\bar{\rho}\frac{\tilde{\varepsilon}^2}{\tilde{k}} \tag{A.11}$$

The terms on the RHS of these transport equations, represent the turbulent transport, the turbulent production and the turbulent dissipation, respectively. The model constants in the standard k,ε model are $\sigma_\varepsilon = 1.3$, $\sigma_k = 1.0$ $c_{\varepsilon1} = 1.44$, $c_{\varepsilon2} = 1.92$.

The deciding benefit of the k,ε-model is its simplicity and cost-efficiency. One main reason for this is the diffusive formulation of the turbulent transports, the first RHS term in Eq. (A.10) and Eq. (A.11). It is in contrast to the convective terms numerically more stable and straightforward to handle (Hallbäck et al., 1995).

A.2   Validation of the Numerical Method: Turbulence Models

The performances of different turbulence models are evaluated via their prediction capability of the cold flow. This is justified because the turbulence modeling in the cold flow is substantially more demanding. The reason for that is the finer turbulence structures as a result of lower viscosity (Sutherland, 1893), and hence higher local Reynolds numbers in the cold flow

$$\mu \sim T^{0.5} \quad \longrightarrow \quad Re = \frac{\rho U \mathcal{L}}{\mu} \quad \xrightarrow[\text{continuity}]{\rho u = const.} \quad Re \sim T^{-0.5} \tag{A.12}$$

The superior turbulence models are subsequently employed to simulate the reacting flow with different combustion models.

In Fig. A.2 the time-averaged and in Fig. A.3 the RMS velocity profiles are presented. These are obtained with the following RANS turbulence models: The standard k,ε of Launder and Sharma

(1974), the Re-Normalization Group k,ε of Yakhot et al. (1992) (RNG-k,ε), the Realizable-k,ε of Shih et al. (1995), the Shear Stress Transport k,ω model of Menter (1994) (SST-k,ω), and the Reynolds Stress Model of Launder et al. (1975) (LRR-RSM).

Similar to the grid independence study, due to the dominance of the resolved large turbulent structures, the results in the first two evaluation sections are very similar. Further downstream from $x = 15mm$, the intermediate and small scale turbulent structures take over, and thus the discrepancies between the RANS models become evident. It can be concluded that the overall prediction accuracy of models increases from the standard k,ε to, Realizable-k,ε, RNG-k,ε, SST-k,ω and LLR-RSM. The computational efforts of all models, except the LRR-RSM, are in the same order. Instead of two equations for k,ε or k,ω, the LLR-RSM solves six equations for each individual Reynolds stresses $\widetilde{u_i'' u_j''}$. Solving the additional transport equations increases the required computational resources by a factor 1.7 higher than the two-equation models. Despite this higher computational demand, LRR-RSM do not exhibit, at least in majority of the flow regions, significant and consistent improvements of the results. Hence, it is disqualified for later parametric design studies.

The streamwise and radial velocities, at the downstream sections $x = 25,35mm$, are predicted more precisely by the SST-k,ω. In contrast to that, the RNG-k,ε exhibit better results for azimuthal velocity components. The peaks of RMS velocities are again better predicted by SST-k,ω, however, on average the RNG-k,ε show similar results. The deciding advantageous of RNG-k,ε was the robustness of the simulations against numerical instabilities in the case of local low mesh qualities and higher CFL numbers. Thus, despite a modest better prediction capability of the SST-k,ω, the RNG-k,ε is selected for the further studies and comparison with LES turbulence models.

As discussed in previous paragraphs, RANS technique is capable of predicting the mean flow quantities, as well as the shape and the location of large scale flow phenomena, in Preccinsta combustor with good overall accuracy. As it can be seen in the Figures A.2 and A.3, RANS has shortcomings, particularly, in predicting the RMS velocities. The small scale velocity fluctuations, however, have a determining role in turbulent combustion as discussed in Sec. 2.2.

LES is a promising tool for predicting the turbulent reactive flows, particularly for dealing with turbulent premixed flames (Gicquel et al., 2008). The main advantageous of LES over RANS is the capability of resolving a great portion of the turbulence spectrum and the associated transient flow phenomena. However, the reaction zone is thinner than the local grid size, and thus, smaller than the resolved turbulence. Hence, the turbulence modeling at the sub-grid level has further a central role in LES of turbulent combustion.

Figures A.4 and A.5 present a comparison of RANS and LES techniques, where the RANS-RNG-k,ε results with medium mesh are plotted against the LES results obtained with the fine mesh and those extracted from the work of Roux et al. (2005). Two LES were performed with different SGS models. The Smagorinsky model with van Driest near-wall damping (Smagorinsky, 1963, van Driest, 1956) exists in the OpenFOAM library, whereas the Singular-Values model (SV) is implemented into the code within this study. It employs the velocity gradient tensor as proposed by Nicoud et al. (2011) to compute the SGS eddy viscosity. Roux et al. (2005) performed a LES

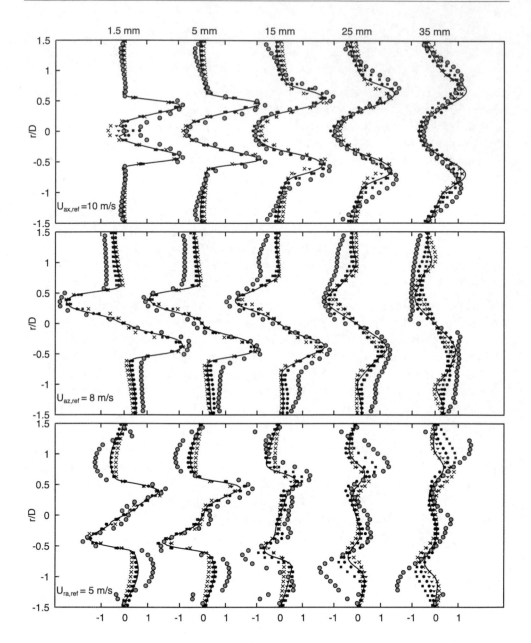

Figure A.2: Nonreacting flow. Time-averaged velocity profiles. Axial (top), azimuthal (middle) and radial (bottom) velocity components are predicted via RANS with Standard k,ε ($\times$), RNG-k,ε (—), Realizable-k,ε ($\blacktriangledown$), SST-k,ω ($\blacksquare$) as well as LRR-RSM ($\bullet$) turbulence models and measured with LDV ($\bigcirc$). The velocity components are normalized with $U_{ax} = 10\,m/s$, $U_{az} = 8\,m/s$, $U_{ra} = 5\,m/s$.

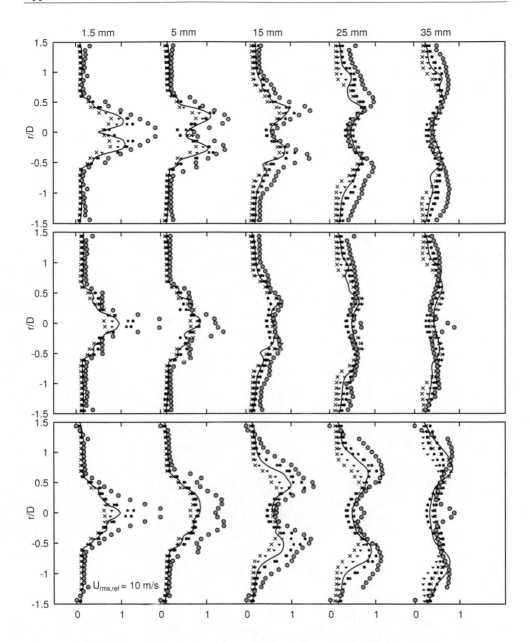

Figure A.3: Nonreacting flow. RMS velocity profiles (modeled ꞁ resolved). Axial (top), azimuthal (middle) and radial (bottom) RMS velocity components are predicted via RANS with Standard k,ε (×), RNG-k,ε (—), Realizable-k,ε (▼), SST-k,ω (■) as well as LRR-RSM (●) turbulence models and measured with LDV (○). The RMS velocity components are normalized with $U_{rms} = 10\,m/s$.

using an in-house CFD code (Colin and Rudgyard, 2000, Moureau et al., 2005). Thereby, the full compressible formulation of the Navier–Stokes equations was solved on a hybrid grid with a third-order accurate discretization of the spatial and temporal terms. They used the WALE-SGS model, which is the predecessor of the SV-SGS model developed by Nicoud and Ducros (1999).

The time-averaged streamwise and radial velocities predicted by RANS and LES with both SGS variants via OpenFOAM are very similar. The LES with WALE model via AVBP code performed better both in predicting the peak values and the shape of velocity profiles. In the case of azimuthal velocities, the LES results of OpenFOAM and AVBP are similar and show, particularly at the two last downstream sections, significant improvements against the RANS results.

For the RMS velocities, the LES results with OpenFOAM exhibit only a marginal improvement compared with the RANS results. It should be noted that in RANS case, especially at downstream sections, up to 80% of the RMS fluctuations were *created* by the turbulence model. Contrary, in LES the RMS signal exclusively consists of the resolved velocity fluctuations obtained from solving the filtered Navier–Stokes equations. Comparing the OpenFOAM-LES results, those with the SV demonstrate a slight improvement over the Smagorinsky model, especially in predicting the velocity fluctuations. For further LES tests, the SV-SGS turbulence model is applied. Owing to its inherent cubic near-wall performance ($\nu_{sgs} \sim y^3$), it does not need an external damping function. Furthermore, the implementation of SV model was straightforward and its computational cost was in same order as the standard Smagorinsky model.

Similar to the mean velocities, the AVBP-LES also better captured the peak RMS values. The main reason for that is the less dissipative nature of the AVBP due to the third-order spatiotemporal discretization of the governing equations. A benchmarking of vortex preservation capability of several LES codes by Jouhaud (2010) confirms this observation as well.

The improved flow prediction by LES is at a higher computational expense. The fine mesh is about eight times larger than the medium one. Furthermore, for a stable and robust LES with OpenFOAM, the CFL numbers of about 0.4 are necessary in contrast to that of 0.8 for RANS. Thus, the time steps for the LES are four times smaller than the RANS, also two times for halved cell dimensions and two times for halved CFL-number. Hence, theoretically, a LES solution of the Preccinsta burner is twenty-four times more expensive than an equivalent RANS. In practice, the LES costs even more because of degressive scalability of the numerical codes performing on a larger number of CPUs. To be specific, for the results presented in Fig. A.4, the RANS required 280 CPUh (1M cells / 60CPUs/ 4.6h) and the LES required 9100 CPUh (8M cells/500CPUs/18.2h), also about thirty-three times higher numerical expense than the RANS. Nevertheless, the reacting flow in Preccinsta model combustor is beside RANS also predicted with LES to have a better assessment of the available numerical tools.

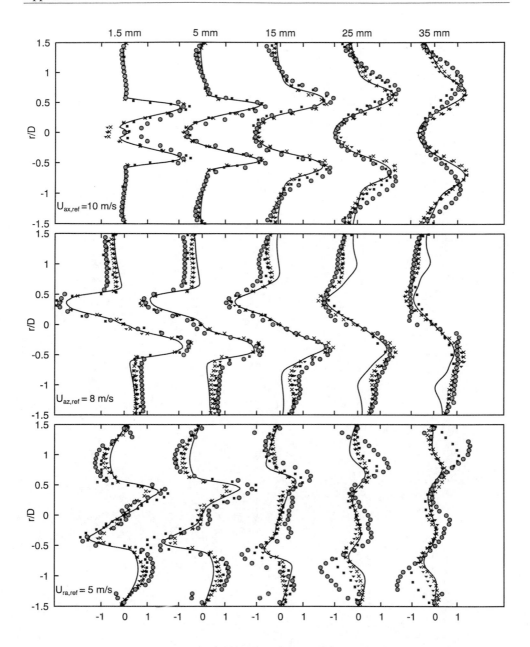

Figure A.4: Nonreacting flow. Time-averaged velocity profiles. Axial (top), azimuthal (middle) and radial (bottom) velocity components are predicted via LES with Smagorinsky (×), Singular Values (▼), WALE-AVBP (■) SGS turbulence models compared with RANS RNG-k,ε (—), and LDV measurements (○). The velocity components are normalized with $U_{ax} = 10\,m/s$, $U_{az} = 8\,m/s$, $U_{ra} = 5\,m/s$.

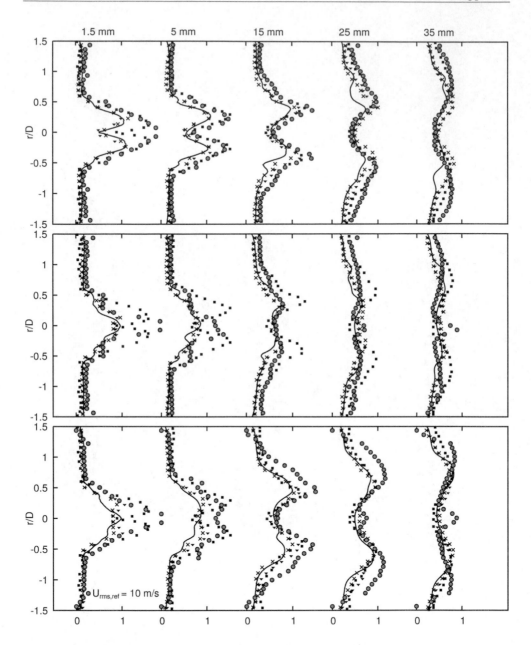

Figure A.5: Nonreacting flow. Resolved RMS velocity profiles. Axial (top), azimuthal (middle) and radial (bottom) RMS velocity components are predicted via LES with Smagorinsky (×), Singular Values (▼), WALE-AVBP (■) SGS turbulence models compared with RANS RNG-k,ε (modeled+resolved —), and LDV measurements (○). The RMS velocity components are normalized with $U_{rms} = 10\,m/s$.

Curriculum Vitae

Name	Behdad Ariatabar
Birthday	19.05.1984
Birthplace	Tehran, Iran

09/2002	School graduation
10/2002-10/2004	Isfahan university of technology, computer engineering
03/2005-09/2005	German pre-university course at RWTH Aachen University
10/2005-06/2012	Diploma in mechanical engineering at RWTH Aachen University Specialization: Energy engineering and turbomachinery Diploma thesis at Siemens AG Subject: Enhancement of Fuel-Air mixing in a novel gas turbine combustor
07/2012-06/2018	Scientific staff and Ph.D. candidate Institut für Thermische Strömungsmaschinen (ITS) (Institute of Thermal Turbomachines) Karlsruhe Institute of Technology (KIT)
10/2018-Present	Principal Engineer Fluid Dynamics Martinrea-Honsel Germany GmbH

Patents

Ariatabar, B, Koch, R, Bauer, HJ (2019): Gasturbinenbrennkammeranordung. Deutsches Patent- und Markenamt. Patent Nr.: DE 10 2017 100 984 B4

Koch, R, Ariatabar, B, Schwitzke, C, Bauer, HJ (2018): Zweistoffdüse. Deutsches Patent- und Markenamt. Dokumenten Referenz-Nr.(DRN): 2018012410533800DE (Pending)